Introduction to Parasitology

Introduction to Parasitology

Grayson Barker

SYRAWOOD
PUBLISHING HOUSE

New York

Published by Syrawood Publishing House,
750 Third Avenue, 9ᵗʰ Floor,
New York, NY 10017, USA
www.syrawoodpublishinghouse.com

Introduction to Parasitology
Grayson Barker

International Standard Book Number: 978-1-64740-013-2 (Hardback)

Cataloging-in-Publication Data

Introduction to parasitology / Grayson Barker.
    p. cm.
Includes bibliographical references and index.
ISBN 978-1-64740-013-2
1. Parasitology. 2. Biology. 3. Molecular parasitology. I. Barker, Grayson.
QL757 .I58 2020
591.785 7--dc23

TABLE OF CONTENTS

Permissions

Index

PREFACE

The study of parasites, their hosts and the relationship between them is known as parasitology. It is a biological field that encompasses principles from diverse disciplines such as cell biology, biochemistry, bioinformatics, immunology, evolution, genetics, molecular biology and, ecology. Parasitology is subdivided into various fields. Some of these are medical parasitology veterinary parasitology, structural parasitology, conservation biology of parasites and parasite ecology. Medical parasitology deals with the parasites that affect humans and the diseases which are caused by them. The study of parasites that infect companion animals and cause economic losses in agriculture is known as veterinary parasitology. Structural parasitology refers to the study of structures of proteins from parasites. This book provides comprehensive insights into the field of parasitology. Some of the diverse topics covered herein address the varied branches that fall under this category. This book will serve as a valuable source of reference for those interested in this field.

To facilitate a deeper understanding of the contents of this book a short introduction of every chapter is written below:

Chapter 1- The organism which lives on or inside a host organism is known as a parasite. It gets its food from the host. The branch of biology which studies parasites, their hosts and their relationship is known as parasitology. This chapter has been carefully written to provide an easy understanding of the varied facets of parasites and parasitology.

Chapter 2- Protozoa are a type of microscopic, one-celled organisms which can be parasitic in nature. Some of the parasitic protozoa are giardia lamblia, entamoeba histolytica, balantidium coli, plasmodium malariae and plasmodium falciparum. The topics elaborated in this chapter will help in gaining a better perspective about these parasitic protozoa.

Chapter 3- A plant which derives some or all of its nutritional requirement from another plant is known as a parasitic plant. Some of the common parasitic plants are Dodder, Rhinanthus minor, Viscum album and Rafflesia Arnoldii. All the diverse aspects of parasitic plants have been carefully analyzed in this chapter.

Chapter 4- Helminths are a type of parasitic worms which feed on a living host in order to survive. Roundworms, tapeworms and hookworms are some of the different types of helminths. Arthropods are a part of numerous parasitic relationships, as both parasites and hosts. This chapter has been carefully written to provide an easy understanding of helminthes and arthropod parasites.

Chapter 5- The life cycle of parasites involves numerous stages where it looks for a host and then infects it in order to derive nutrition. The study of host-parasite interactions plays an integral role in exploring the biological processes governing parasitic diseases. The topics elaborated in this chapter will help in gaining a better perspective about the life cycles of parasites and the host-parasite interactions.

Finally, I would like to thank the entire team involved in the inception of this book for their valuable time and contribution. This book would not have been possible without their efforts. I would also like to thank my friends and family for their constant support.

Grayson Barker

Chapter 1

Understanding Parasitology

The organism which lives on or inside a host organism is known as a parasite. It gets its food from the host. The branch of biology which studies parasites, their hosts and their relationship is known as parasitology. This chapter has been carefully written to provide an easy understanding of the varied facets of parasites and parasitology.

Parasites

A parasite is an organism that lives within or on a host. The host is another organism. The parasite uses the host's resources to fuel its life cycle. It uses the host's resources to maintain itself.

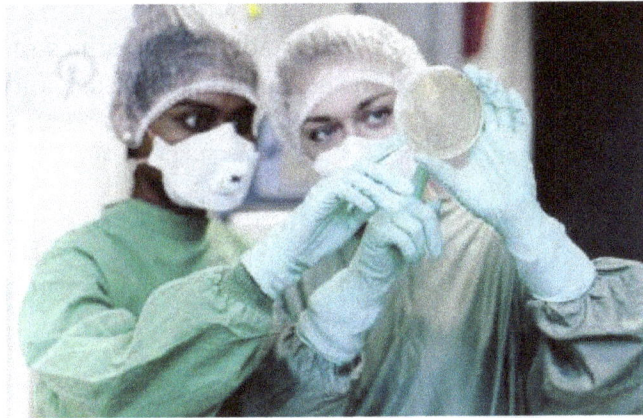

Parasites range from microscopic in size to over 30 meters in length.

Parasites vary widely. Around 70 percent are not visible to the human eye, such as the malarial parasite, but some worm parasites can reach over 30 meters in length.

Parasites are not a disease, but they can spread diseases. Different parasites have different effects.

Endoparasite

These live inside the host. They include heartworm, tapeworm, and flatworms. An intercellular parasite lives in the spaces within the host's body, within the host's cells. They include bacteria and viruses.

Endoparasites rely on a third organism, known as the vector, or carrier. The vector transmits the endoparasite to the host. The mosquito is a vector for many parasites, including the protozoan known as Plasmodium, which causes malaria.

Epiparasite

These feed on other parasites in a relationship known as hyperparasitism. A flea lives on a dog, but the flea may have a protozoan in its digestive tract. The protozoan is the hyperparasite.

Types

There are three main types of parasites:

1. Protozoa: Examples include the single-celled organism known as Plasmodium. A protozoa can only multiply, or divide, within the host.

2. Helminths: These are worm parasites. Schistosomiasis is caused by a helminth. Other examples include roundworm, pinworm, trichina spiralis, tapeworm, and fluke.

3. Ectoparasites: These live on, rather than in their hosts. They include lice and fleas.

Worms

Worms, or helminth organisms, can affect humans and animals.

1. Anisakiasis: This is caused by worms that can invade the intestines or the stomach wall. The worms are passed on through contaminated fresh or undercooked fish and squid.

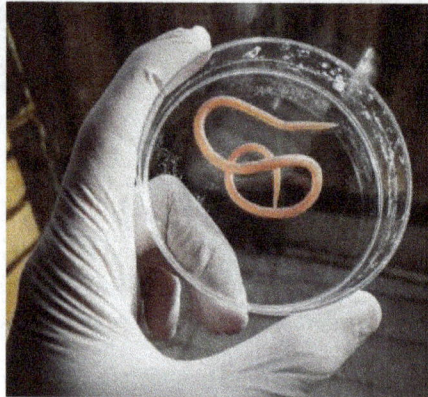

Roundworms can be passed on by raccoons.

2. Roundworm: Ascariasis, or a roundworm infection, does not usually cause symptoms, but the worm may be visible in feces. It enters the body through consuming contaminated food or drink.

Raccoon roundworm: Baylisascaris is passed on through raccoon stools. It can affect the brain, lungs, liver, and intestines. It occurs in North America. People are advised not to keep raccoons as pets for this reason.

3. Clonorchiasis: Also known as Chinese liver fluke disease, this affects the gall bladder. Humans can become infected after ingesting raw or poorly processed or preserved freshwater fish.

Dioctophyme renalis infection: The giant kidney worm can move through the wall of the stomach to the liver and eventually the kidney. Humans can become infected after eating the eggs of the parasite in raw or undercooked freshwater fish.

4. Diphyllobothriasis tapeworm: This affects the intestines and blood. Humans can become infected after eating raw fish that live wholly or partly in fresh water. Prevalence has increased in some parts of the developed world, possibly due to the growing popularity of sushi, salted fillets, ceviche, and other raw-fish dishes.

5. Guinea worm: This affects subcutaneous tissues and muscle and causes blisters and ulcers. The worm may be visible in the blister. As the worms are shed or removed, they enter the soil or water, and are passed on from there.

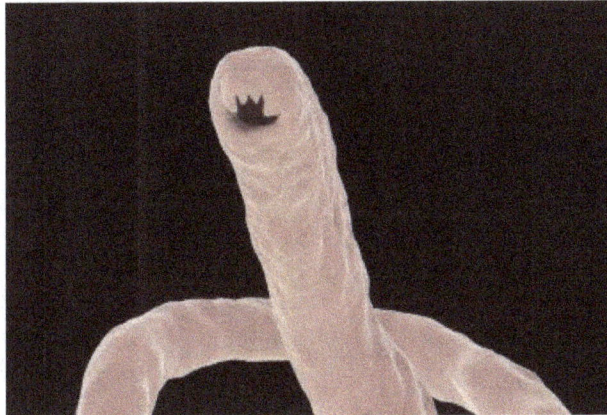
Hookworms can cause intestinal disease.

6. Hookworm: These can cause intestinal disease. They lay their eggs in soil and the larvae can penetrate the skin of humans. Early symptoms include itching and a rash. They are most common in damp places with poor sanitation.

7. Hymenolepiasis: Humans can become infectedby ingesting material contaminated by rodents, cockroaches, mealworms, and flour beetles.

8. Echinococcosis tapeworm: Cystic echinococcosis can lead to cysts in the liver and lungs, and alveolar echinococcosis can cause a tumor in the liver. Humans can be infected after eating foods contaminated by the feces of an infected animal, or from direct contact with an animal.

9. Enterobiasis pinworm: A pinworm, or threadworm, *Enterobius vermicularis* can live in the colonand rectum of humans. The worm lays eggs around the anus while a person sleeps, leading to itching. It spreads through the oral-fecal route.

10. Fasciolosis liver fluke: This affects the gall bladder and liver. It is common in countries where cattle or sheep are reared, but rare in the U.S. It can affect the liver and the bile ducts and it causes gastrointestinal symptoms. It passes from one mammal to another through snails. A person may get it from eating watercress, for example.

11. Fasciolopsiasis intestinal fluke: This affects the intestines. It can also transmitted when consuming contaminated water plants or water.

12. Gnathostomiasis: This causes swellings under the skin, and occasionally affects the liver, the eyes, and the nervous system. It is rare, but it can be fatal. It occurs in Southeast Asia. It is transmitted by eating freshwater fish, pigs, snails, frogs, and chicken.

13. Loa loa filariasis: Also known as loaisis, this is caused by the *Loa loa* worm, or African eye worm. It causes itchy swellings on the body. It occurs mainly in Central and West Africa and is transmitted through deerfly bites.

14. Mansonellosis: This is passed on through the bites of midges or blackflies. It affects the layers under the surface of the skin, but it can enter the blood. It can lead to angioedema, swellings, skin rash, fever, and joint problems. It is present in Africa and Cental America.

15. River blindness: Caused by a worm known as *Onchocerca volvulus*, this affects the eyes, skin, and other body tissues. It is found near fast flowing water. It is transmitted through the bite of a blackfly. It occurs in South America, but 90 percent of cases are in Africa.

16. Lung fluke: Also known as paragonimiasis, this affects the lungs, causing symptoms similar to those of tuberculosis (TB). However, it can reach the central nervous system, leading to meningitis. It is transmitted when eating undercooked or raw freshwater crabs, crayfishes, and other crustaceans. It is most common in parts of Asia.

17. Schistosomiasis, bilharzia, or snail fever: There are different types of schistosomiasis. They can affect the skin and internal organs. It results from exposure to fresh water that has snails in it that are infected with the blood fluke, or trematode worm. The worms are not found in the U.S. but they are common worldwide.

18. Sparganosis: Humans can become infected if they eat foods tainted with dog or cat feces that contains the larvae of a tapeworm of the Spirometra family. It can lead to a migrating abscess under the skin. It is rare.

19. Strongyloidiasis: This can lead to severe and possibly fatal immunodeficiency. The parasite penetrates through the skin and affects the lungs, skin, and intestines. It is passed on through direct contact with contaminated soil. It most occurs in tropical and subtropical regions.

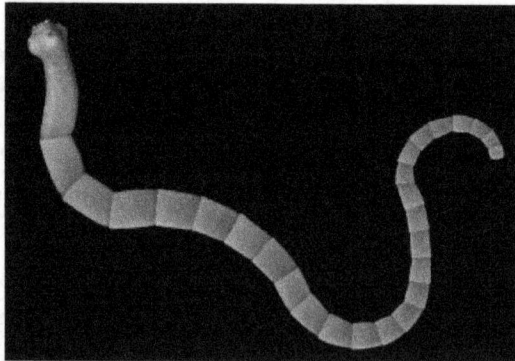

Different types of tapeworm can affect the intestines, the liver, or the lungs.

20. Beef and pork tapeworms: Taeniasis is caused by tapeworms of the taenia family. They affect the intestines. They are passed on by eating undercooked beef or pork.

21. Toxocariasis: A roundworm transmits this infection from animals to humans. It affects the eyes, brain, and liver. It is caused by accidentally swallowing the eggs of the parasite, for example, when young children play with soil. Nearly 14 percent of people in the U.S. have antibodies, suggesting that millions have been exposed. Most never have symptoms.

22. Trichinosis: This is caused by the roundworm of the Trichinella family. Infection can lead to intestinal symptoms, fever, and muscle aches. It is passed on by eating undercooked meat.

23. Whipworm: Also known as trichuriasis, whipworms live in the large intestine. Eggs are passed in feces. It is common all over the world. Humans can become infected when ingesting the eggs, for example on unwashed fruit or vegetables.

24. Elephantiasis lymphatic filariasis: This is transmitted through mosquito bites. The adult worms live in the lymph system. Infection can lead to lyphedema and elephantiasis, in which swelling can cause disfigurement and disability. In the Americas, it is passed on by the *Culex quinquefasciatus* mosquito.

Ringworm is sometimes mistaken for a worm, but it is not a worm. It is a fungal infection.

Ectoparasites

These are parasites that live on the outside of the body, such as fleas.

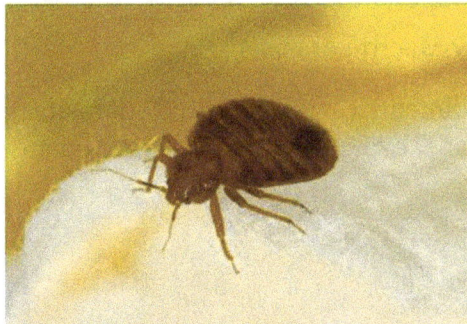

Bed bugs are ectoparasites: They live on the outside of the body.

1. Bedbug: These can affect the skin and vision. They are found all over the world. Sharing clothing and bedding can spread infection. They may be present in newly rented accommodation and hotel rooms.

2. Body lice: These are common worldwide. Infection can spread through sexual activity, skin-to-skin contact, and sharing bedding or clothing.

3. Crab lice: These affect the pubic area and eyelashes. They are common all over the world and spread through sexual activity, skin-to-skin contact, and sharing bedding or clothing.

4. Demodex: These affect the eyebrow and eyelashes. They are common all over the world and can spread through prolonged skin contact.

5. Scabies: This affects the skin. It is common all over the world and can spread through sexual activity, skin-to-skin contact, and sharing bedding or clothing.

6. Screwworm: This is transmitted by a fly, and it affects skin and wounds. It is found in Central America and North Africa.

7. Head lice: These live on the scalp and affect the hair follicles. They are common all over the world and spread through head-to-head contact. A reaction to their saliva causes itching.

Parasites come in many shapes and sizes and can lead to a wide variety of symptoms and health issues. Some parasites are treatable and others are not.

Macro-Parasites

Macro-parasites are multicellular, eukaryotic organisms that are large enough to be seen with the naked eye. Macro-parasites, like other parasites are metabolically dependent on other living organisms, referred to as the host organism. Most parasites grow inside the host but generally reproduce by infective stages outside of the host.

Type of Macro-parasites

There are many species of macro-parasites. The most common of these include nematodes, ticks, mites and flatworms. Macro-parasites can be either classed as endoparasites; parasites that live inside the host, or ectoparasites; parasites that live on the host. Examples of endoparasites include flukes and tapeworms, while examples of ectoparasites include mosquitoes, fleas, ticks, leeches and lice.

Examples of Diseases caused by Macro-parasites

There are many diseases and ailments caused by macro-parasites as well as micro-parasites that affect plants, animals and humans. Examples of diseases in humans that are caused by macro-parasites include:

- Fleas
- Ticks
- Tapeworm
- Bilharzia worms
- Hydatid worms
- Liver fluke.

Parasitology

Parasitology is a study of relationships; more specifically, it is the scientific study of the biologic relationship known as parasitism.

A close relationship between two living organisms is known as symbiosis. Organisms involved together in symbiosis are referred to as symbionts or symbiotes. Symbiotic relationships range from being very superficial (simple) to extremely intimate (complex). Symbiotic relationships may be as simple as one symbiont being transported by another; such a transportation-based relationship is referred to as phoresy. A simple symbiotic association may evolve into a more complex physiologic dependence of one species on another species. At least one of the two symbionts will receive some measurable benefit from the relationship. When one symbiont receives benefit from the other, but causes no harm or produces no measurable benefit, the relationship is referred to as commensalism. If both symbionts receive some benefit from the relationship, it is known as mutualism. Parasitism is the term describing relationships where one symbiont, the parasite, receives some benefit from the relationship, while the other symbiont, the host, is harmed.

The terminology used in describing different symbiotic relationships seems very precise when defined in a text, yet the boundaries of these associations can be blurred in "the real world". Commensalism may exist while the host's nutrition is adequate or as long as the host has a properly-functioning immune system. Most authorities agree that morbidity and mortality constitute harm, thus illness and death as consequences of a physiologic relationship between two organisms are evidence of an obvious parasitic relationship. Upsets in some of the delicate balances necessary to maintain commensalism between two organisms can result in benefit for one organism in the relationship at the expense of the other, *i.e.*, parasitism. A liberal definition of harm may be necessary to accept some relationships as belonging under the heading of "parasitism". Some parasites inflict obvious structural and/or metabolic damage, while many others may injure by simply increasing the energy needs of the host. A host may need to increase food intake to compensate for the effects of a parasite. Cells and proteins will be directed toward maintaining an immune response against a parasite, expending resources that could be needed to fight a more serious pathogen. Increased needs for nutrients may force the host to modify its behavior patterns in order to obtain more food, thus exposing the host to other potential threats. Parasitism over extended periods of time can be an expensive circumstance for the host.

The harm to hosts caused by parasites can be grouped into seven basic categories:

1. Parasites can inflict mechanical damage, with their physical activities (movement, feeding, mating, *etc.*) directly altering the normal architecture of the host cells/tissues/organs/systems.

2. Parasites can compete with the host for vital substances.

3. Parasites can release/excrete toxic chemicals that potentially have deleterious effects on the host.

4. Parasites can trigger dangerous inflammatory/immune responses that can damage the host's own tissues.

5. Parasites can put pressure on host energy utilization and reserves.

6. Parasites can cause pain.

7. Parasites can provoke adverse changes in host behavior.

Parasites that feed on skin or live most of their lives in association with the integument of the host, thus producing an infestation, are known as ectoparasites. Parasites that dwell primarily in the internal tissues of the host, thus producing an infection, are known as endoparasites.

Parasites and hosts are individuals living in large, complex populations, communities and ecosystems that include many other organisms. The parasite and the host interact with each other, but they must interact with the other species (predators, scavengers, etc.) around them, too. Adaptation is the key to parasite success. The adult parasite (or the most well-differentiated form of the parasitic organism) feeds and reproduces sexually in or on an organism specifically referred to as the definitive host. Success of the parasitic species depends on, among other things, the success of its various host species. Some parasites develop a relationship with only one definitive host species; such parasites are described as being very host-specific. Other parasites are not very host-specific; they are able to proliferate and reproduce sexually in several species of definitive hosts. Parasitism is full of risks for both the parasite and the host. If the host becomes extinct, the extremely host-specific parasite will probably disappear as well. A parasite that has adapted the ability to survive in several definitive hosts increases its chances for success, over evolutionary time, in at least one species.

Survival of a parasite requires that the parasitic species must be able to spread from one susceptible host to another susceptible host. Transmission stages are forms of parasites that, when in appropriate contact with a susceptible host, are capable of survival and/or development in such hosts. Parasites often produce, in definitive hosts, transmission stages (eggs, larvae, cysts, etc.) that are released into the bloodstream, intestinal contents, urine, *etc.*; these stages exit the definitive host and are intended to survive the threats posed by the external environment, for the purpose of infecting another definitive host. Some transmission stages are protected from environmental extremes of temperature and humidity only by structural design; upon release from the definitive host, or perhaps after some required time outside of the definitive host have passed, these transmission stages can passively or actively infect another definitive host. Parasites that spread directly from definitive host to definitive host have a direct life cycle. Some parasites have adapted to life by producing transmission stages that live, develop, and persist in other organisms in the surrounding community. The parasite does not undergo sexual reproduction in these organisms, but required development, or transformation from one life cycle stage to another, occurs.

Organisms that are required for the development of immature or asexual stages of parasites are known as intermediate hosts. Asexual reproduction is possible with certain parasites in their respective intermediate hosts. Intermediate hosts become infected with immature or asexual stages of parasites by ingesting transmission stages, such as eggs, larvae, and cysts, produced by parasites in the definitive host, by ingestion of post-transmission stages inhabiting other intermediate hosts, *etc.* Definitive hosts acquire parasite transmission stages by eating various types of infected intermediate hosts, or by being fed on by hematophagous arthropod intermediate hosts. Other organisms, known as paratenic hosts, may harbor transmission stages of parasites acquired in various ways; paratenic hosts however arenot required for the completion of the parasite's life cycle. Parasites do not undergo substantial change in form inside a paratenic host, although some parasites may increase in size in a paratenic host. In the known parasite-paratenic host relationships that have been studied, the parasitic stages can persist for long periods of time in the paratenic host. Parasites may be transmitted between intermediate and paratenic hosts, and likewise transmission may occur between multiple paratenic hosts. Paratenesis is one way that parasite transmission stages can greatly extend their lifespans in an ecosystem. Paratenic hosts are not physiologically required by a parasite, but ecologically they may be very important for the regular completion of the parasite's life cycle. Parasites that use intermediate and paratenic hosts have indirect life cycles.

Parasites succeed, biologically-speaking, by surviving and reproducing. In a perfect world (from the parasite's perspective), parasites adapt to environmental conditions and/or they infect definitive hosts, intermediate hosts, and paratenic hosts. In a complex ecosystem, some of these randomly-distributed parasite transmission stages may take paths that will not lead to success. Many transmission stages succumb to unsuitable environmental extremes of temperature, humidity, exposure to sunlight, *etc.* Parasites may enter aberrant hosts, in whom conditions are not adequate for survival, further development, or sexual maturity. Aberrant hosts are a dead-end for both the individual parasite and the parasite species. Almost every occurrence of a parasite in an aberrant host results in the failure of aberrant parasites to either survive or to produce viable transmission stages.

These are a few of the most basic concepts of parasitology. There are exceptions to any rules, and the definitions used in parasitology are constantly tested by new discoveries about parasites, hosts, and parasitic relationships.

Parasitism

Parasitism is an association or a situation in which two organisms of different taxonomic positions live together where one enjoys all sorts of benefits (like derivation of nourishment, reproduction etc. which are basic requirements for existence) at the expense of the other. The benefited organism is called the parasite and the organism harbouring the parasite is called the host.

Hosts are not hospitable to parasites. Instead they consider parasites as foreign bodies and want to exterminate or overpower them by operating various devices like: producing antibodies, increased peristalsis, diarrhoea, mucus secretion, encystation by host tissues etc. Parasites to avoid host's reaction for existence develop many specialities like increased fecundity, polyembryony, safe-habitat, production of special enzymes, a good deal of transmission etc.

Due to close contact/intimate association, the responsive reactions and resistance displayed by a host to its parasite and the protective devices adopted by a parasite in response to its host's reactions in order to establish them in their respective environments are called host-parasite-interactions. Parasitism is a very broad term and different types of parasites are recognised on different basis.

Basis of classification	Types of parasites	Definition and example
Duration	Temporary or Partial	Visits its host for a short period, e.g. adult dog flea.
	Permanent or Total	Leads a parasitic life throughout the whole period of its life e.g., life cycle of Trichina worm.
Degree of dependence	Facultative or Optional	Lives a parasitic life when opportunity arises e.g. Ancylostoma.
	Obligatory or Compulsory	Cannot exist without a parasitic life, e.g. Taenia solium.
Position	Ectoparasites or Ectozoa	Lives outside on the surface of the body of the host e.g. lice.
	Endoparasites or Endozoa	Lives inside the body of the host, i.e. in blood or in digestive tract, e.g. Taenia saginata.

In the course of their life cycle, parasite may become associated with more than one host. In many cases the life cycle is characterised by numerous very rigid requirements. Whenever a parasite is able to live and reproduce within a host—the result is an elaborate host- parasite interactions.

Host Specificity of Parasites

In mature condition a given parasite is quite often found in limited number of hosts. In extreme condition, distribution of a parasite may be restricted to a single host—mono-specific parasite. Even when poly-specific the different hosts are phylogenetically related. This host specificity is a function of physiological specialization and evolutionary age.

It is broadly divided into two parts:

(a) Ecological specificity: The parasites are capable of making room in a foreign host but normally never reach another host due to ecological barriers. Such parasites are able to develop in more host-species under laboratory conditions than in nature.

(b) Physiological specificity: The parasites are physiologically incapable of surviving and reproducing in a foreign host, e.g., Taenia solium in dog survives but never achieves reproductive ability. If the parasites find the conditions suitable for their development- then it is said to be compatible with that of the host. If not, it is said to be incompatible.

Host and Parasite - A Mutual Relation

In the course of time a mutual adjustment or relation or tolerance frequently develops between the two which permits them to live together as a sort of compound organization without very serious effect or damage to either.

The virulant types, however, try to eradicate the hosts. But it is essential to keep the host alive and not to kill it by causing a high degree of pathogenecity. By killing the host it will ultimately lead to death of itself also. Accordingly Natural Selection leads to the elimination of most virulent species and maintains the less virulant ones.

Effects of Parasites on Hosts

The effects of parasitism on the hosts are intimately associated to the effect of host on the parasites. These effects depend on several factors, such as-age, diet, genetic factors, susceptibility of the hosts, the size, number and virulence of the parasites, their mortality, migration, and method of feeding.

Destruction of Host's Tissues

Time of Injury

(a) Some parasites injure the host's tissue during the process of entry, e.g. hookworms like Ancylostoma duodenale, whose infective larvae inflict extensive damage to cells and underlying connective tissue while penetrating the host's skin.

(b) Some inflict tissue damage after they have entered, e.g., larvae of Ascaris lumbricoides while passing through lungs of human host cause physical damage to lung-tissue, leading to pneumonia.

(c) Others induce to histopathology changes by eliciting cellular immunologic response to their presence, e.g. Entamoeba histolytica actively lyses the epithelial cells lining the host's large intestine and liver causing large ulcerations by the action of secreted enzymes.

Types of Cell Damage

Three major types:

- Parenchymatous or albuminous degenerative cells become swollen and packed with albuminous or fatty granules and pale cytoplasm. This type of damage is characteristic of liver, cardiac muscle and kidney cells.

- Fatty degeneration cells are filled with an abnormal amount of fat deposits, e.g. liver cells.

- Necrosis means any type of persistent cell degeneration which finally die, e.g. as the result of encystment of Trichinella spiralis in mammalian skeletal muscles; necrosis of surrounding tissue is followed by calcification.

Tissue Changes

Four main types:

(a) Hyperplasia:

- Refers to an increased rate of cell division resulting from an increased level of cell metabolism.

- Leads to a greater total number of cells but not in their sizes.

- This commonly follows an inflammation and is the consequence of an excessive level of tissue repair.

- For example—thickening of bile duct in presence of Fasciola sp. is the result of hyperplasia.

(b) Hypertrophy:

- Refers to an increase in cell size.

- Commonly associated with intracellular parasites.

- For example in Erythrocytic phase of Plasmodium vivax, the parasitized RBC's are commonly enlarged. Spermatogonial cells of Polymnia nebulosum (an Annelid) when parasitized with Caryotropha mesnili (a Protozoan), are enlarged.

(c) Metaplasia:

- Refers to the changing of one type of tissue into another without the intervention of embryonic tissue.

- The encapsulating epithelial cells and fibroblasts of the fluke, Paragonimus westermani in human lungs are transformation of certain other type of cells in the lungs.

(d) Neoplasia:

- This is the growth of cells in a tissue to form a new structure, e.g., a tumour.

- Neoplastic tumour is not inflammatory.

- This is not required for the repair of organs.

- It does not conform to a normal growth pattern.

- It may be benign or malignant.

Example: Eimeria sp. causes tumor in rabbit liver, Schistoma mansoni in human intestine and liver, Echinococcus granulosus in human lungs etc.

Competition for Host's Nutrients

- Endoparasites with a great density causes nutritional deficiency in host by absorbing sugars, vitamins, amino-acids etc.

- Mal-nourished hosts are more proned to disease and infection.

Example: Diphyllobothrium latum (a fish tapeworm) in human causes anaemia by absorbing profuse Vitamin B_{12} (as much as 10 to 50 times more than do other tape-worms). Vitamin B_{12} plays an important role in blood formation, thus its uptake by D. latum results in anaemia.

Utilisation of Host's Non-nutritional Materials

Parasites in some cases also feed on host- substances, other than stored or recently acquired nutrients. Ectoparasites and endoparasites feed on host's blood, 500 human hookworms can cause a loss of about 250 ml blood/day, leading to anaemia.

Mechanical Interferences

Mechanical interferences by parasite cause injuries to hosts, e.g. elephantiasis or filariasis in humans is caused by Wuchereria bancrofti. Increased number of those adult worms in lymph vessels coupled with aggregation of connective tissue may result in complete blockage of lymph flow.

Excess fluid behind the blockage seeps through the walls of lymph ducts into the surrounding tissues, causing edema and ultimately with scar tissues—the elephantiasis of limbs, breasts, scrotum etc.

Effects of Toxins, Poisons and Secretions

Specific poisons or toxins egested, secreted or excreted by parasites cause irritation and damage to hosts, e.g.

- Antienzymes produced by intestinal parasites counteract host's digestion.

- Allergin, a toxin as the body fluid of nematodes—Parascaris equorum and other ascarids, irritates the host's cornea and nasopharyngeal mucous membrane.

- Toxin of pathogenic Entamoeba histolytica produces toxic symptoms in parasitized mammalian hosts and creates ulcerations within the large gut of man.

- Schistosome cercarial dermatitis is the result of an allergin reaction against an irritating parasitic secretion from the fluke.

- Haemozoin pigments produced by trophozoites of P. vivax exert toxic effect in infected persons and the patients suffer from periodic effect of high fever with chilliness and shivering.

Other Parasite - Induced Alterations

(a) Sex reversals:

Gonads of parasitized hosts may change, leading to sex reversals; e.g. crab when parasitized by Sacculina (a crustacean) display sex reversals. Parasitized male crab acquired secondary female characteristics like broad abdomen, appendages modified to grasp eggs, chelae become smaller, testes with testicular cells at various stages of degeneration.

Parasite-removed male develops into hermaphrodite by regeneration of rest testicular cells. Parasitized female crab shows ovarian degeneration but does not show hermaphroditism on removal of parasite, as ovarian tissue cannot regenerate.

(b) Parasitic castration:

- It refers to destruction of host's gonadal tissues by a parasite.

- Reduces egg and/sperm production in host's body or becomes sterile.

- The mudflat snail—Ilyanassa obsoleta are directly castrated by the trematode— Zoogonus lasius. Sporocysts of Z. lasius secrete a molecule that causes the destruction of host reproductive cells as well as inhibits same to genesis.

The freshwater snail, Lymnaea stagnalis is indirectly castrated by larvae (Sporocysts) of Trichobilharzia ocellata (a trematode). These larvae do not possess mouth and thus destroy the gonadal tissue by chemical means.

(c) Enhanced growth of host: An interesting aspect of parasite induced change in hosts is responsible for enhanced growth; e.g.

- Workers of the ant, Pheidole commutula become much larger when parasitized by the nematode, Mermis sp.

- Fresh water snail, Lymnaea ariculata infected with trematode larvae is larger than uninfected ones.

- Mice infected with larvae of Spirometra mansonoides (a tapeworm) grows faster than non-parasitized one.

- Rats when parasitized by Trypanosoma lewise increase their weight more rapidly than non-parasitized one.

The enhanced growth of the host is due to stimulation of growth-promoting molecules secreted by the parasites.

Host Reaction

In immuno-parasitology, the animal is the host and the parasite is either self (by molecular memory) or non-self (foreign).

When a host recognizes the parasite as non-self, it generally reacts against the invader in two ways:

1. Cellular (or cell mediated) reactions: Where specialised cells become mobilised to arrest and eventually destroy the parasite as usual.

2. Humoral reactions: Where specialised molecules in circulatory system (antibodies/immuno-globulin's in case of vertebrates) interact with the parasite, usually resulting in its immobilization and destruction.

Internal Defense Mechanisms

The internal defense mechanisms of animals, both invertebrates and vertebrates, are of two types:

- Innate (or natural)

- Acquired.

Theoretically each of them again can be of two types—cellular and humoral.

Invertebrate Immunity

Innate Internal Defense Mechanism

Cellular Factors

These includes the following chief categories:

(a) Phagocytosis:

When a foreign parasite (small enough to be phagocytosed) invades into an invertebrate host, it is usually phagocytosed by the host's leucocytes, primarily the granulocytes.

Phagocytosis consists of three phases:

- Attraction of phagocytes to the non-self material, commonly by chaemo-taxis.

- Attachment of foreign material to the surface of the phagocyte, usually involving a specific chemical binding site.

- Internalization of the foreign substance i.e. engulfment by the phagocyte.

Fate of phagocytosed parasites:

- May be degraded intracellularly.

- May be transported by phagocytes across epithelial borders to the exterior.

- May remain undamaged within the phagocytes and some may even multiply within host cells.

(b) Encapsulation:

- Parasites, that are too large to be phagocytosed, are encapsulated as invading non-self mass enveloped by cells and/or fibres of host origin, as found in insect and molluscan hosts.

- Encapsulation consists of:

  ◦ First leucocytosis (increase in number of leucocytes);

  ◦ Migration—many of these cells migrate by the process of chaemo-tactic movement to-wards the parasite and form a capsule of discrete cells around it, as found in insects and in other cases (i.e., in molluscs).

Host cells synthesize fibrous material which becomes deposited inter-cellularly and concentrically in layers around the parasite.

- Encapsulation of Tetragonocephalum (a tapeworm) in the American oyster, Crassostrea virginica. Fate of encapsulated parasite: Destroyed and disintegrated parasite's tissues are phagocytosed by host's granulocytes.

(c) Nacrezation (i.e. pearl formation):

- Nacrezation is another type of cellular defense mechanism, known in molluscs.

- As certain helminth parasites, e.g., Meiogymnophallus minutus (a trematode) occurs be-tween the inner surface of the shell (nacreous layer) and the mantle of marine bivalves. Now the mantle is stimulated to secrete nacre that becomes deposited around the parasite. In so doing, a pearl is formed and the enclosed parasite is killed.

(d) Melanization:

- The process involves deposition of the black-brown pigment, melanin around the invading parasite.

- Melanization is chemically the result of enzymatic oxidation of polyphenol by tyrosinase.

- This is detrimental to the parasite and may lead to its death by interfering with such vital activities like hatching, moulting or feeding.

- Melanization of the nematode, Heterotylenchus autumnalis in haemocoel of larval house-fly—Musca domestica.

Humoral Factors

These fall into two categories:

(a) Innate humoral factors:

These are two types:

- Those are directly parasitocidal, e.g. several marine molluscan species contain a constituent in their tissue extract that is lethal to Cercariae of the trematode, Himasthla quissetensis.

- Those that enhance cellular reactions, e.g., naturally occurring agglutinins or lectins. These glycoprotein molecules enhance phagocytosis of the non-self-material.

(b) Acquired humoral factors:

These are also of two types:

Lysosomal enzyme: When challenged with non-self-parasites, some invertebrate's granulocytes (haemocytes) hypersynthe size certain lysosomal enzymes and subsequently release them into some parasites. When they come in contact with elevated enzymes are killed either directly or indirectly whereas the parasite's body surface by action of lysosomal enzyme undergoes chemical alteration and thus is recognized as non-self-material and consequently get attacked by host's haemocytes.

Antimicrobial molecules: When challenged with micro-organisms, some insects synthesize antimicrobial molecules which are quite different from vertebrate antibody but kill the microorganisms, e.g.; the synthesis of two small basic proteins (PgA and PgB) by the moth, Hyalophora cearopia when challenged with E. coli and these proteins kill the bacterium.

Vertebrate Immunity

Immunity refers to resistance against disease caused by a foreign agent. This is based on antigen-antibody interaction. In vertebrates this reaction is very specific. Antigen is the only foreign substance (Proteins, glycoproteins, nucleoproteins etc.) which on introduction induces the synthesis of antibody under some appropriate conditions. All zoo-parasites theoretically contain multiple antigens.

These are chiefly of two types:

- Somatic antigens molecules comprising some of parasites.

- Metabolic antigen molecules are associated with secretion and excretion, e.g., moulting fluid of nematode is highly antigenic.

Antibody—Refers to proteins synthesized by host tissue in response to the administration of an antigen and which specifically react with that antigen to immobilize and destroy it.

Mechanism of Antigen-antibody Interactions

- Antigens on introduction are able to bind with specific cell surface receptor of lymphocytes (both B and T-lymphocytes).

- Host lymphocytes are now stimulated to proliferate and differentiate.

- As a consequence, clones of progeny lymphocytes are formed.

- In the process of proliferation, some progeny differentiate into effector cells (the functional end products of the immune response). Plasma cells are B-lymphocyte effector cells that secrete antibodies. Killer T-cells are such T-lymphocyte effector cells that eliminate foreign cells simply by contact.

- As soon as immunoglobulin's are produced, immunogens are coated with such antibodies and are rapidly destructed and / or phagocytosed.

The parasites try to establish it within the host while the latter tries to destroy it which results in dynamic state of equilibrium. The reaction of the host in the presence of a parasite is termed as resistance. If resistance is sufficiently high to prevent parasite reproduction, it is known as absolute resistance and if parasite is able to overcome it and still reproducing it is called partial resistance.

(a) In case of larger parasites there is considerable damage to host tissues where histamin is released, macrophages are attracted and a primary stage of inflammation is set up.

(b) In the second stage, the cells of the lymphoid macrophage system elaborate antibodies. The immunoglobulin's appears in various molecular forms differing in properties and actions. The macroglobulin's (IgM) is the first to appear in an infected animal. This is followed by the appearance of gamma-globulin and alpha-globulin (IgA). The properties and number of antibodies vary from individual to individual parasitic infections.

(c) Interferons also play an important role in the immunity reaction of the host. These are known to operate in malaria and other viral reactions by rendering the host cells unfit for habitation by intercellular parasites.

Categories of Antigen-antibody Interactions

These are of three types:

1. Primary interaction: Refers to the basic event during which the antigen is bound to and/or more available sites on the antibody molecule.

2. Secondary interactions: Include agglutination, precipitation, complement-dependent reaction, neutralisation, immobilisation etc.

- Agglutination reaction: Antibodies (agglutinins) clump microbes representing antigens and visible conglomerates are formed. This is referred to as agglutination reaction.

- Lysin and lysis reaction: Lysin (antibodies) dissolves or lyses antigens. The reaction occurs in the presence of complement, a substance in normal serum representing a system of enzymes. Complement is sensitive to heat, chemical substances, ultraviolet rays, long-term storage etc.

- Complement-dependent reaction: In the first phase of this reaction mutual adsorption of antigen-antibody takes place and precipitation occurs. In the second phase of reaction the fixation of complement by antigen-antibody occurs which is used for detecting many of the parasitological infections.

- Precipitin reaction: Precipitin is the antibody that brings about the formation of a minute deposit (precipitation) when interacted with specific antigen (precipitinogen). While in agglutination the entire microbial bodies act as antigen, in precipitin the antigen will be the results of breakdown of microbial bodies or their products. This precipitin reaction is used for detecting infections like plague, anthrax, tularaemia etc.

- Phagocytosis: Refers to the engulfment of non-self-material by host cell like macrophages. In vertebrates, this is introduced through the action of antigen, antibody and complement.

This occurs in two ways:

- Accumulation of leucocytes through complement sequence and

- Certain antibodies called opsonins become coated on to the foreign materials and they enhance phagocytosis.

Opsonins are antibodies occurring in normal as well as in immune sera which inhibit microbes making them more amenable to phagocytosis.

3. Tertiary interaction: Refers to in vivo expressions of antigen-antibody reactions. At times these may be of survival value to the host, but at other times they may lead' to a disease through immunologic injury.

Immunity to Parasites

Vertebrate hosts always develop some degree of acquired immunity in the presence of parasites.

This is usually of two types:

- Concomitant immunity: Where immunity, either complete or partial, may be maintained only while the parasites are present.

- Sterile immunity: Where immunity persists long after the complete disappearance of the parasites.

Protozoan Blood Parasites

- Cattle infected with certain species of Babesia shows premonition.

- While cattle, long after Theileria parva has disappeared, shows sterile immunity.

- For malaria-causing parasites—Premonition usually occurs in case of avian- infecting species of Plasmodium while sterile immunity can be produced with the rodent malaria causing agents P. berghei and P. vinckei.

- Certain breeds of cattle develop partial immunity against Trypanosoma.

Protozoan Tissue Parasites

(a) Antibodies are produced by host cells when they come in contact with antigens of parasites. Thus,

- When Entamoeba histolytica resides in host's intestinal lumina, no detectable antibody occurs, but when the Amoeba invades the mucosa and other tissues, antibodies are evident.

- For Leishmania tropica, if the sores produced by this parasite are allowed to heal spontaneously, the host becomes totally immuned to reinfection.

- Immunity to Toxoplasma gondii is evident in adult humans who show the antibodies, while congenital infection with T. gondii leads usually to death and those who survive, show a hydrocephalus condition. Immunity can control the intracellular stages only if the macrophage becomes successfully activated, which requires production of cytokines by T-cells.

- The trophozoites of gregarines (Gregarina sp) exhibit an eerie gliding movement. Various theories have been put forward. Similar movement has been noticed for the trophozoites of Plasmodium and Eimeria. There is little evidence of an effective immunity against merozoites of Eimeria sp and Gregarina sp.

Experimental studies have shown that immunity can directly affect intracellular developmental stages in both initial and subsequent infections. This immunity relies on the production of cytokine interferon, presumably activating an intracellular defence mechanism that is capable of preventing sporozoite and merozoite development.

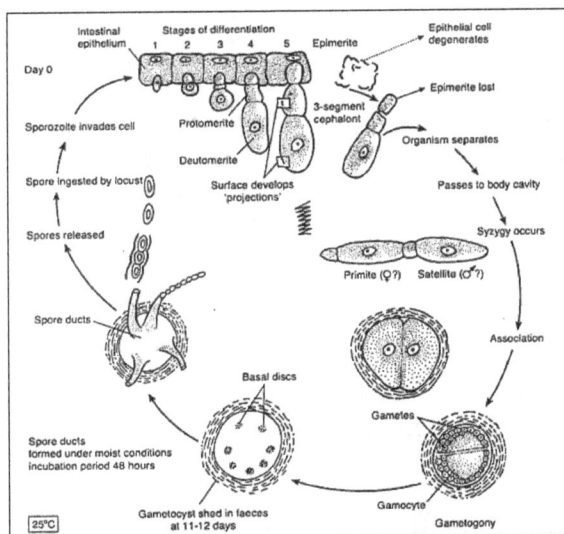

Diagrammatic representation of the life cycle of Gregarina garnhami in the desert locust, "spores'are sometimes referred to as oocysts or sporocysts.

(b) Special manifestations are shown by Trypanosoma: Parasites may change their surface antigens during their life cycle in vertebrate hosts. Continuous variation in major surface antigens is also shown by African trypanosomes. In such case, infected individual show waves of blood parasites. Each wave comprises of parasites expressing a surface antigen that is different from the previous wave.

Thus, by the time the host produces antibodies against the parasites, an antigenically new organism has grown out. Such continuous antigenic variations make it difficult to effectively vaccinate the infected individuals.

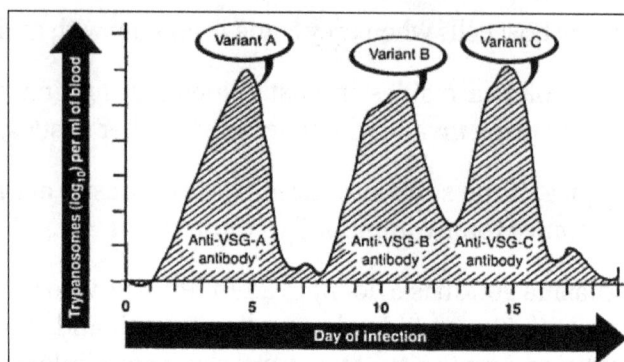

Antigenic variation in African trypanosomes

Blood Helminths

- Include adult Schistosomes and certain microfilariae. Concomitant immunity is shown by rodents on Rhesus monkeys against infection of Schistosoma.

- Microfilariae of Wuchereria, Brugia etc. can survive long period in human blood, indicating in their normal hosts they invoke very little immune response.

Tissue Helminths

Example: Trichinella spiralis, Echinococcus granulosus etc. produce antibodies against tissue nematode.

The serious and often fatal results of trichiniasis are due to the offspring of the infecting worms and not to the adult worms in the intestine. The resistance of mice to Trichinella infection can be enhanced by injection of secretions and excretions of Trichinella larvae and adults. T. spiralis arctica is essentially incapable of infecting rats but readily infects mice, deer and certain carnivores.

- Hosts can inhibit totally or partially the establishment of the parasitic development of Cooperia curticei in sheep.

- Parasites develop anatomic abnormalities along with oedema and necrosis, decreased albumin levels and significant reduction in weight gain and often death follows; e.g. Ostertagia sp. It fails to develop a vulvar flop in calves.

- Parasite does not grow as rapidly or reach its normal size, e.g., Trichinella spiralis in man.

- Parasite's life-cycle is altered, e.g. Strongyloides sp., in naive pigs etc.

Antibodies, although, do not generally cause the death of adult worms but affected worms undergo retarded growth with destrobilisation.

Protective Immunity is absent in case of tissue trematodes like Fasciola sp., Paragonimus westermani etc.

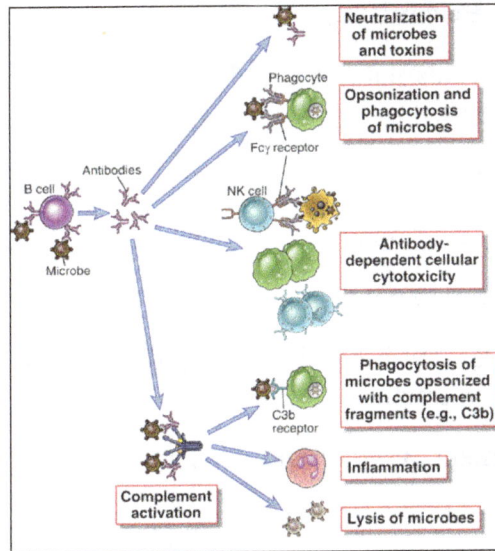

Molecular Mimicry

Endoparasites, both protozoans and helminths can survive long durations in immunologically hostile environments. Such parasites are immunologically inert because they produce immunogens which are antigenic, similar to those of the hosts so that they are recognised as 'self'.

This phenomena of antigen shearing between hosts and parasites is called molecular mimicry.

There are three possibilities regarding its origin:

- Mimicry by natural selection,

- Mimicry by host induction, and

- Mimicry resulting from incorporated host antigens.

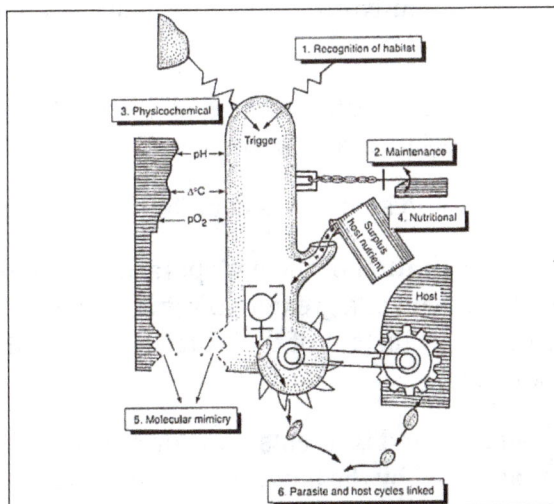

This is a cartoon-like representation of the 'ideal parasite' which: 1. 'Recognises' a host site: 2. maintains its position; 3. is adapted to the physico-chemical conditions of the host; 4. utilises host nutrient in a manner compatible with host survival; 5. presents a surface with molecular configuration; 6. has a life cycle synchronised with that of the host

Effects of Parasitism on Parasites

The parasites undergo elaborate adaptations in order to harmonize with the host. This adaptation takes place by two ways as follows:

(a) Degeneration of organ system:

- Sense organs are poorly developed.

- Less developed locomotory organs.

- Intestinal parasites have no well-developed digestive organs, e.g., in Taenia solium there is complete loss of digestive tract.

(b) Specialization: New attainment of organs or systems

In External Parasites

- Compressed bodies with backwardly projecting spines for attachment, e.g., Flea.

- Tactile hairs in mites.

- Barbed proboscis in ticks.

The ectoparasites also show degeneration in the following organs:

- Loss of eyes and sometimes sense organs.

- Loss of wings e.g., Flea.

- Sometimes reduction of legs.

In Internal Parasites

The endoparasites possess peculiar admixture of both degenerated and specialised parts as follows:

- They possess all sorts of hooks, barbs, suckers as organs for attachment, although they have lost sense organs or special organs for locomotion.

- Very simple nervous system and sometimes complete loss of digestive systems.

- The most remarkable specialization of the endoparasites brought forth in their reproductive systems which enabled them to a greater power of reproduction. For them every structure, every function, and every instant of their life are modified to a certain extent for the sole purpose of reproduction.

"A fluke does not eat to live"—it lives and thus eats only for reproduction. The highest specialization of reproductive system is achieved by the tapeworm. In the mature proglottids both male and female reproductive organs are present in tapeworm while others (proglottids) contain either male or female reproductive systems.

Host Susceptibility and Specificity

For a parasite to live habitually in a host-body there must be:

- Suitable conditions for the transmission of parasites from one host to another;

- Able to establish itself in a host when it reaches a new one;

- Satisfactory conditions for growth and reproduction after its establishment.

Hosts and Parasites: A Mutual Tolerance

In course of time, a mutual adjustment or tolerance frequently develops between the two - host and parasite, which results to lead their life as a sort of compound organisms, without any serious effect to either. It may not be in the best interest of the parasites to destroy its host for it would do it—invariably it would destroy itself.

Thus, parasitism is a specialised mode of living within a broader ecological category— symbiosis. In parasitism—the host and the parasite have a very intimate association where all the benefits are derived by the parasites from its prey—the host and the two systems constantly interacting with each other.

Thus the criticism of the story—"Host- Parasite Interaction" is one of the compromise—a key (parasite) to unlock the box (host) of an unrevelled mysterious entity—in which the parasites making elaborate efforts to overcome the match against the host while the host making attempts to keep the ball in the goal of parasites, thus trying to eradicate it.

Host

Hosts are the animals which lodge parasites. They are larger than the parasites in size.

- Definitive host: The definitive host is the one in which a parasite reaches sexual maturity and undergoes reproduction. It is mostly a vertebrate.

- Intermediate host: The host in which some development of the parasite occurs but it does not reach sexual maturity, is referred as intermediate host and is usually a invertebrate.

- Paratenic or transport host: Sometimes the parasite enters a host in which it does not undergo any development but remains alive till it gains entry in the definitve host or intermediate host. Such a host is termed as paratenic or transport host or a carrier host. These hosts are important for the completion of the life cycle of certain parasites as they are believed to bridge the ecological gap between the intermediate and the definitive host.

- Reservoir host: An animal which is utilized by a parasite as a temporary refuge till it reaches its appropriate host is referred as reservoir host. It is generally a animal which is normally infected with a parasite that can also infect man. For example, dogs and cats are reservoirs of *Leishmania*.

All the parasites follow one principle as far as the selection of the host is concerned. Some parasites are specialized to a single species of the host where they live and multiply, while others have become adapted to a wide variety of hosts. For example, a malarial parasite is found in man as well as in birds but the species of malarial parasite found inman is not the same as the one found in the birds. This is called as host- specificity.

Another notable feature of the parasite is that, the parasite can infect different hosts but within each host it can survive in definite tissues or organs only. This can be explained by the example of malarial parasite which is found in different hosts like man and birds but in either of these hosts the tissue which it has selected is same i.e. blood. The parasites are, thus, not only host- specific but tissue- specific also.

The parasite causes damage to the host and for every parasite to act as a disease agent there has to be its source or reservoir. It must also adopt a definite system to gain entry and thrive well in the body of the host in order to derive maximum benefit from the host.

The parasite can only be called successful if it gets all the advantages from the host without endangering life of the host because the parasite itself would have no future if the host dies. The manner in which the parasite makes its way in the body of the host and the extent to which it can cause damage is often studied under the heading of dynamics of disease transmission.

Parasitic Nutrition

Parasites derive nutrition from plants or animals without killing them. This is called parasitic nutritive strategy.

Example: Cuscuta, Lice, Leaches, Tape worms.

The type of nutrition in which an organism obtains its food from other organisms is called parasitic nutrition. Here the organism which obtains the food is called "parasite" and the organism from which food is absorbed is called "host". Parasites live in intimate association with host from which it derives its essential material and gives no benefit to the host in return. Host may be either a plant or an animal.

Examples: Several fungi, several bacteria, a few higher plants like Cuscuta. Some animals like Plasmodium and Round worm.

Evolutionary Biology of Parasitism

Ecological and Evolutionary Concepts

The following concepts are grouped as a start to the synthesis of evolutionary biology and parasitology. The concepts are necessarily basic in order to retain generality, but they may form the ground work for development of more sophisticated concepts. The general approach will compare the ecology and evolution of parasites with that of predators.

Parasites are adapted to Exploit Small, Discontinuous Environments

For a parasite each host exists in a matrix of inhospitable environment. For very small organisms a wide dispersion of resources makes colonizatiop of new hosts hazardous. Adaptations to reduce this hazard include:

- Mass production of spores or eggs (e.g. tapeworms);

- Dispersal of inseminated females which form a high proportion of the population;

- Dispersal by phoresy making host discovery accurate.

In each of these cases a single female can found a colony on a new resource remote from its origin. Multiplication in the colonizer's progeny may lead to a new and relatively isolated population. New propagules colonize other isolated resources. Thus parasites tend to exist in small homogenous populations with little gene flow between them.

Types of reproduction in parasites which permit a single female to found a new population include inbreeding among progeny, hermaphroditism (e.g. trematodes, cestodes), or asexual reproduction (polyembryony in digenetic trematodes and parasitic wasps, parthenogenesis in many taxa). The commonness of parthenogenesis among parasitic arthropods is not generally appreciated. Arrhenotoky (the production of males from unfertilized eggs) occurs in all Hymenoptera, probably the majority of Thysanoptera, many iceryine and aleurodid Homoptera, and in some Coleoptera. It is a major mode of reproduction in certain families of mites. Thelytoky (the production of females from unfertilized eggs) occurs sporadically in a great variety of insects and cyclical parthenogenesis) is prevalent in aphids (Aphididae) and gall wasps (Cvnipidae), and it occurs in gallflies (Cecidomyiidae)and many parasites in other taxa. Thus parthenogenesis occurs in five of the ten largest families of parasitic insects on plants in Britain: Cecidomyiidae, Curculionidae, Aphididae, Tenthredinidae and Cynipidae. In the animal parasites of the British fauna eight of the largest families have all species parthenogenetic since they are within the order Hymenoptera: Ichneumonidae, Braconidae, Pteromalidae, Eulophidae, Lamprotatidae, Platygasteridae, Encyrtidae, and Diapriidae.

Table: Number of species in the ten largest families in the British insect fauna in each of the categories listed. Primary parasites feed on plants and secondary parasites feed on animals. Families marked with an asterisk contain some or all members reproducing through parthenogenesis. The family Cynipidae contains some secondary parasites.

Predators		Primary Parasites		Secondary Parasites	
Dytiscidae	110	Cicidomyiidae*	629	Ichneumonidae*	1938
Sphecidae*	104	Curculionidae*	509	Braconidae*	891
Coccinellidae	45	Aphididae*	365	Pteromalidae*	493
Corixidae	32	Tenthredinidae*	358	Eulophidae*	485
Cucujidae	32	Noctuidae	298	Tachinidae	228

Hemerobiidae	29	Chrysomelidae	248	Philopteridae	176
Vespidae*	27	Cicadellidae	242	Lamprotatidae*	156
Asilidae	26	Cynipidae*	238	Platygasteridae*	147
Anthocoridae	25	Olethreutidae	216	Encyrtidae*	144
Soldidae	20	Miridae	186'	Diapriidae*	125
MEAN	45	MEAN	329	MEAN	478

Parthenogenesis is so common in parasites and yet its adaptive nature for parasites is not well understood. Some adaptive features follow:

1. A single female can establish a new colony, and multiplication can occur when the probability of contact between more than one individual of the same species is very low.

2. Parthenogenetic parasites may seem to be maladapted to a patchy, complex and changing environment since progeny of a single female are likely to be more uniform than progeny from females producing fertilized eggs. However, this disadvantage can be reduced by utilization of highly stable and predictable microenvironments provided by the homeostasis of living organisms. A positive feedback may reinforce the evolution of parthenogenesis in parasitic organisms.

3. The gross reproductive rate of thelytokous females is doubled, so the probability of finding a new host is similarly increased.

4. Particularly adaptive combinations of genes are fixed in perpetuity in thelytokous multiplication: automictic thelytoky tends to produce homozygous individuals while apomictic thelytoky involves mitotic divisions only.

This locking of gene combinations may be especially important in parasitic organisms where large banks of genes are likely to be involved with close coevolutionary tracking of the host system. Disruption of such a block would generate gross maladaptions with almost certain lethal results.

Parthenogenesis is relatively rare in predatory taxa such as Odonata, Heteroptera and spiders. Of the 10 largest families of predators in the British insect fauna only the two hymenopterous families, Sphecidae and Vespidae, show any form of parthenogenesis.

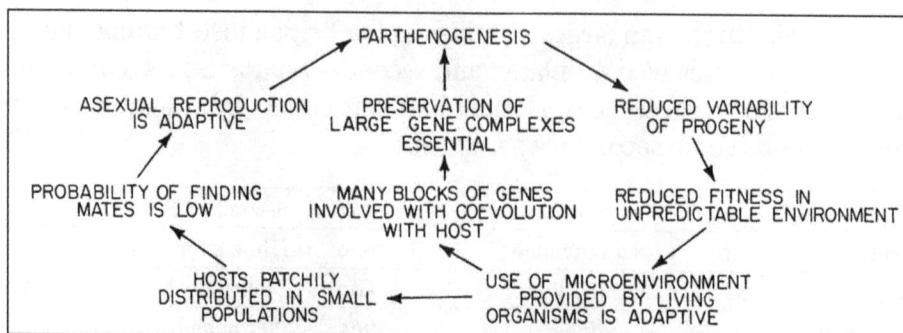

The positive feedback loop which may promote the evolution of parthenogenesis among parasitic organisms or at least permit it to exist. It is assumed that parthenogenetic species may revert back to sexual reproduction under favorable selection pressure.

Parasites represent the extreme in the exploitation of coarsegrained environments.

The small size of parasites, the specific cues used in discovery of relatively large food items, and the poor mobility of many stages in the life cycle means that they must exploit a coarsegrained environment in the sense of MacArthur and Levins and MacArthur and Wilson. Such small organisms, which face relatively large habitat differences compared to the tolerance of individuals may respond in two ways to environmental variability. If the environment is stable in time and variable in space, Levins predicts that the population should be monomorphic and specialized with a geographic pattern of discrete races. If the environment is uniform in space and variable in time the population should be polymorphic, with specialized types, and have a geographic pattern of clines in frequencies of these specialized types. Since hosts probably represent a variable resource both in time and space the development of both geographic races and polymorphism in parasites may be common, although races will be more frequent than polymorphism with the population structure as envisioned here. Specialization permits a relatively large number of species to pack into a given set of resources so many species may coexist under equilibrium conditions.

In contrast to parasites, predators are relatively large and mobile and exploit a relatively fine-grained environment. Habitat differences are small compared to the tolerance of the animals. Levins predicted that such populations will be monomorphic and unspecialized, with geographic patterns showing continuous clines. Few of these generalists can coexist under equilibrium conditions.

Evolutionary Rates and Speciation Rates are High

The environmental constraints on the parasitic habit described in Concepts 1 and 2 promote the fractionation of gene pools and produce an advantage to inbreeding or asexual reproduction. They foster rapid divergence of populations, race formation, and eventual speciation. Short life cycles, short generation times and high fecundity result in high reproductive rates which permit dramatic changes in population size and rapid differentiation of populations under dissimilar selective regimes. Predominantly homozygous populations (or haploid males in parasitic Hymenoptera) have all genes exposed to selection in each generation. Both the founder effect and genetic drift are probably significant in speciation of parasites. Both Wright and Carson have explored the evolutionary potential of populations with this structure and they concluded that the probability of evolution and speciation was much higher than in a randomly breeding population of the same size.

Evidence for high evolutionary rates may be seen in the many sibling species, subspecies, host races, and different types observed in parasitic organisms. Mayr states that sibling species are especially common among insects and singles out the Lepidoptera, Diptera, Coleoptera and Orthoptera as providing abundant examples. In the Coleoptera he notes that sibling species are particularly common in the Curculionidae, Chrysomelidae, and Cerambycidae, three of the largest families in the order with the majority of species parasitic on plants. Zimmerman suggests that five or more new species of Hedylepta (Lepidoptera) must have evolved within 1000 years in Hawaii since they are endemic and specific to banana which was only introduced that long ago. Roelofs et al. and Klun et al. describe the sibling species of the corn borer which respond to different isomers of the otherwise identical pheromones. Bush states that a new race of the western cherry fruit fly (Rhagoletis indifferens) was extant on domestic cherry within 89 years of the plants' introduction to Northwest North America. Jones discusses the several types of leishmaniasis caused by biologically distinct forms within each of the species Leishmania donovani and L. tropica. He also suggests on theoretical grounds that speciation can occur in one generation, a view held by Bush based on studies of herbivorous parasites.

The large size of many families of parasitic insects also supports the concept of high evolutionary and speciation rates. Even though some parasitic taxa evolved much later than some predatory taxa, families of parasites on plants are on average almost eight times larger than those of predators, and families of parasites on animals are over ten times larger. In both groups of parasites the tenth family is larger than the first ranked family of predators.

Adaptive radiation is extensive and its degree of development in each taxon of parasites depends upon:

a) The diversity of hosts in the taxon or taxa being exploited (i.e., numbers of species and the degrees of difference between species).

b) The size of the host target available to potential colonizers (body size, population size, geographical distribution).

c) The evolutionary time available for colonization of hosts.

d) The selective pressure for coevolutionary modification (i.e., for specialization).

e) The mobility of hosts.

To understand the evolution of the large numbers of parasite species part of the answer can be found in an understanding of why there are so many resources available and what influences the presence or absence of parasites on these resources.

Diversity of Hosts

Adaptive radiation can be most extensive when many related species of host are available for colonization, particularly if the hosts within a taxon differ in an important way relative to the requirements of the parasites. Many examples could be given of a fairly weak relationship between size of family of hosts and numbers of parasites that exploit members of that family. For example, there are more of the leaf mining flies in the family Agromyzidae on large families of plants than small families. However, the regression accounts for only 61% of the variation and much of the remaining scatter is probably accounted for by the chemical diversity of plants in each family. Where species in a family are chemically distinct, for example in secondary metabolic products, or as Hering noted, in plant proteins, parasites are predominantly monophagous (i.e., feed on host species within a single genus). Parasites utilizing hosts in families of low chemical diversity tend to be oligophagous (i.e., feed on species in several genera). Monophagous species have a narrow feeding niche; therefore many species can pack onto a given number of host species relative to oligophagous parasites. Only the family Graminae is an exception to this pattern perhaps because of the exceptional commonness of species and individuals making colonization more probable. The example of the Agromyzidae supports Eichler's rule, well known to parasitologists, which states"When a large taxonomic group (e.g., family) of hosts consisting of wide varieties of species is compared with an equivalent taxonomic group consisting of few representatives, the larger group has the greater diversity of parasitic fauna."

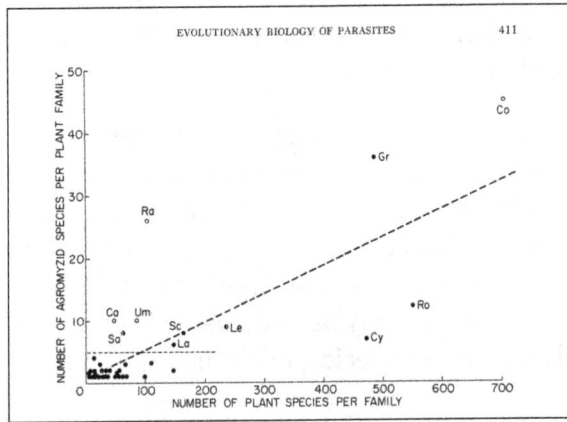

Relationship between number of species in each plant family and the number of
agromyzid flies attacking each family in Canada and Alaska.

In figure, for those plant families attacked by more than five agromyzid species (above small dashed line) the following symbols are used: (open circle) agromyzids monophagous, (closed circle) agromyzids oligophagous, (closed|open) some monophagous and other oligophagous, (open|closed) only two genera in family. Sa-Salicaceae, Ca-Caprifoliaceae, Um-Umbelliferae, Ra-Ranunculaceae, Sc-Scrophulariaceae, La-Labiatae, Le-Leguminoseae, Gr-Graminae, Cy-Cyperaceae, Ro-Rosaceae, Co-Compositae. The regression line is described by the formula $Y = 0.7648 + 0.0455 \times$ ($r2 = 0.61$, $P < .01$).

At each successive trophic level the diversity of potential hosts' increases and the opportunities for adaptive radiation expand, within the limits discussed below.

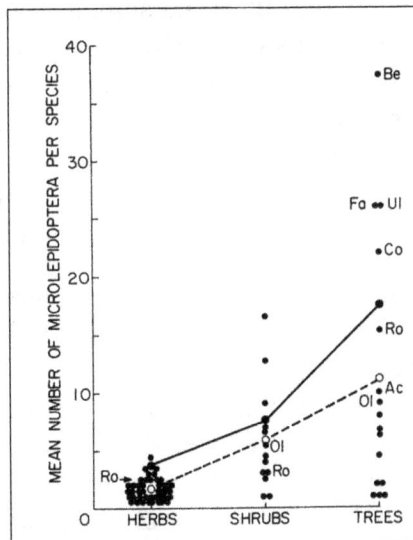

Relationship between plant form of a family and the average number of
microlepidoptera per plant species attacked in each family.

In figure, host species from Ford. Ol-Oleaceae, RoRosaceae, Be-Betulaceae, Fa-Fagaceae, Ul-Ulmaceae, Co-Corylaceae, Ac-Aceraceae. Open circles = the mean number for each plant form.

Dashed line = the trend of increased parasites per plant species as plant size increases. Mean number of leafmining Lepidoptera per plant genus for each plant form = large closed circles. Solid line = trend per plant genus. Host species from Hering Families with representatives in more than one plant form were subdivided, as shown for the Oleaceae and Rosaceae. The large range in numbers of species on trees is discussed in Concept 4c.

Thus, primary parasites (parasitic herbivores) may be abundant, with many large families included, but radiation has been even more extensive in the largest families of secondary parasites (parasitic carnivores). The family Ichneumonidae is over three times larger than the largest family of primary parasites in the British fauna, and Townes estimates a total of about 60,000 species must exist in the world. Only at the fourth trophic level in a parasite food chain (e.g., the obligate hyperparasites of parasitoids) resources may be so dispersed and hard to find that a trend towards generalization, and reduced numbers of species, might be expected.

The resulting diversity of parasitic organisms supported after the evolution of a new genus or species may be impressive. An extreme case is seen in the genus Quercus where even in the small fauna of Britain the oak supports at least 390 species of parasitic herbivores. The number of secondary parasites supported by these herbivores must be even higher. The winter moth, Operophtera brumata, is known to support 63 species of parasite and microlepidoptera in the genus Phyllonorycter may support about 50 species of parasitic insects.

Size of Host Target

The large number of parasites which coexist on oaks is possible partly because of the tree's large size. The many resources available for colonization and the species that exploit them have been described by Morris. The average numbers of microlepidoptera per species in a family of plants are much less limited in trees than in shrubs and much less in shrubs than in herbs. Leaf mining Lepidoptera, Coleoptera and Hymenoptera illustrate similar trends when Hering's keys are analyzed. A similar response may be observed among ectoparasites (e.g., biting lice, Mallophaga, and feather ticks, Sarcoptiformes) of birds, with large birds having more regions of the body which are sufficiently distinct so that different parasites can colonize and maintain a competitive edge in each region. For example, in an extensive survey Foster discovered three species of mallophagan which occupied two regions of the body on the orange-crowned warbler, Vermivora celata, whereas Dubinin found seven species on the larger Ibis falcinellus located in four distinct body regions.

Host population size as well as the geographic distribution of a species are equally important in determining the probability of a parasite reaching a new host, colonizing it and maintaining a lasting relationship with the host. Large populations and extensive range of a potential host must increase the chances of colonization by parasites. Perhaps Kellogg was the first to regard hosts as islands. Janzen also points out that hosts can be regarded as islands available for colonization by parasites and they are therefore subject to the insights of the theory of island biogeography, as Dritschilo et al. have demonstrated for mite species on cricetid rodents in North America. Probability of colonization is influenced by island size which can be regarded as host size, host population size or range size.

Evolutionary Time Available

In the previous section trees were shown to support on average more parasitic herbivores than herbs. However, the range in average numbers of parasites per species of tree is extreme and is not accounted for by the size of the host individuals. Most of this range can be explained by the relative evolutionary opportunity provided by each species of tree. Common trees which have existed in a region over considerable spans of time have high numbers of parasites whereas recently available hosts with a restricted range have small numbers.

Once parasites have colonized a host, divergence of host stock and eventual speciation leads to divergence of the parasites and, depending on the time involved, results in host-race formation or new parasite species. Thus parasites can be extremely useful in elucidating the phylogenetic relationships of their hosts. Such diverse groups as beetles on pines, termitophiles in termite colonies, and ciliates on turbellarians have provided clues to inconspicuous differences between hosts. This concept has been formalized into a rule by Fahrenholz: common ancestors of present-day parasites were themselves parasites of the common ancestors of present-day hosts. Degrees of relationship between modern parasites thus provide clues as to the parentage of modern hosts. The more specific Fuhrman's rule, that each order of birds has its particular cestode fauna, implies a similar relationship between the evolution of host and parasite.

Selective Pressure for Co-evolutionary Modification

The importance of interaction between host and parasite in the evolution of both has been stressed by numerous authors. The stepwise co-evolutionary process results in extreme specialization and complex defense mechanisms. As described in Concepts 2 and 3 specialization is likely to increase the rate of speciation which may occur in both host and parasite. Indeed, as Atsatt points out, parasites may have increased the adaptive potential of their angiosperm hosts enabling the evolution of heterotrophic species, including the parasitic flowering plants.

The importance of co-evolutionary pressure was illustrated in the relationship between agromyzid leaf miners and plants. Families containing species which are biochemically distinct had relatively high numbers of agromyzids which were also specialists. This is seen particularly clearly when host ranges are compared for agromyzid leaf miners on Graminae and Umbelliferae. The latter family is composed of aromatic plants which produce a diverse array of essential oils and related resins with a large number of pharmaceutically interesting species. Although the Umbelliferae is a much smaller family than the Graminae many more agromyzid species attack members of the family in Europe (61 species on Umbelliferae, 35 species on Graminae; from analysis of keys by Hering. This is apparently because the chemical diversity of potential hosts within the Umbelliferae has forced specialization of the parasites and 82% of the species of agromyzid attack only one genus each; they are monophagous.In contrast only 29% of agromyzids are monophagous on grasses. In addition, there are fewer agromyzids on each genus of Umbelliferae than of Graminae. No more than seven species occur on any one genus in the former family and 15 genera have only one parasite species. When there are few parasite species per host, coevolution can proceed rapidly since adaptive reactions need not be compromised by conflicting adaptations in response to other parasites exerting different selective pressures. Thus adaptations for specialization are reinforced, isolation of populations becomes more likely, and speciation is more rapid. As Mayr expresses it, "Host specificity is thus an ideal prerequisite for rapid speciation. The end result is a comparatively large number of specialists attacking the Umbelliferae whereas many more generalists attack the Graminae.

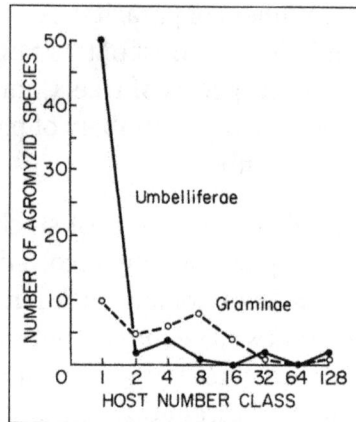

Frequency distribution of the number of genera attacked by agromyzid parasites in the host plant families
Graminae and Umbelliferae. Host number classes on a logarithmic scale with the first number
in the class given. Thus class 8-15 = 8.

The degree to which specialization is demanded is a potent force in adaptive radiation. Szidat's
Rule states that the more specialized the host group, the more specialized are its parasites; and,
conversely, the more primitive or more generalized the host, the less specialized are its parasites.
Hence, the degree of specialization may serve as a clue to the relative phylogenetic ages of the
hosts. Predators must remain generalized and radiation has been unimpressive. Plant parasites
show varying degrees of radiation depending on the intimacy of their association with the host.
Plant bugs (Miridae) are mobile, relatively large ectoparasites, although immature stages spend
much time on a single host. In the British fauna 186 species have been identified.

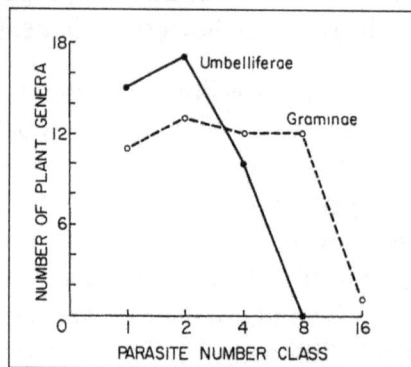

Frequency distribution of the number of parasitic agromyzids per genus of plants in the families Graminae
and Umbelliferae. Parasite number classes arranged on a logarithmic scale.

The much smaller plant lice (Aphididae) are more sessile, more intimately associated with the host
and have undergone more extensive radiation (365 species in British fauna). Some are gall-mak-
ers. The largest families of plant parasites are predominantly endoparasitic. The weevils (Curculionidae) as larvae mine in leaves, under bark, in shoots or roots, or feed in rolled leaves, fruits and
seeds. They number 509 species. The most highly coevolved parasitism occurs in the gallforming
endoparasitic flies, Cecidomyiidae, which are also the most numerous (629 species) primary parasites in the British fauna. Eastop referring to aphids noted that the production of a distinctive gall
is always associated with host-plant specificity. Members of the Cynipidae (238 species) also form
galls but radiation has not been so extensive; evidently the opportunities for radiation have not
been as fully exploited as by the gall flies.

Specialization in parasites on animals should be more highly developed than in parasites on plants. Although animals as a group are chemically more similar than plants, herbivores have a greater diversity of places to live than plants, and strong and diverse behavioral, phagocytic and immune defenses against parasites. Thus, finding and living with hosts seems to demand a greater number of adaptations per species for parasites of animals than for parasites of plants, and therefore a narrower host range. Members of the largest families in the British fauna are parasites of insects: the Ichneumonidae (1938 species) and Braconidae (891 species).

Probably the degree to which species of host and parasite are coevolved and the proportion of the genome devoted to coadaptation could not have been appreciated without extensive breeding of plants for parasite resistance. Flor and Day provided strong evidence from plant breeding experiments that there is a gene-for-gene relationship between the plant host's resistance and the associated parasite's virulence. After prolonged coevolution, where the stepwise process has escalated defenses many times, many such complementary gene-for-gene pairs must exist between parasite and host. This complex of genes coadapted to counterparts in the host, like the closed variability system described by Carson, must be maintained or fitness is reduced drastically. These blocks of genes may be maintained by inversions, development of supergenes, or by cloning, the last so often seen in parasitic organisms.

Mobility of Hosts

The mobility of hosts may either dampen or promote the evolutionary process towards speciation of parasites. High mobility of host stages which are infected with parasites reduces isolation of parasite populations important in speciation of parasites. Static hosts such as plants or hosts of low mobility reinforce isolation of parasite populations. Plants disperse as seeds, a stage usually not infected by parasites (important exceptions are fungi, seed wasps and seed weevils). Similarly, insects tend to disperse as very early larval stages before parasites attack or more commonly in the adult stage which is relatively free from parasitic insects. The contrary situation exists among some mammals, birds and fish, which are relatively large, highly mobile animals. Parasitized animals disperse, gene flow between parasite populations is high and rates of divergence and speciation are reduced. It is among the students of parasites on these groups that the concept of slow evolution of parasites, relative to that of their hosts, seems to have originated. Jordan supported this view based on studies of fleas (Siphonaptera) which are restricted to birds and mammals, as are the lice (Phthiraptera) studied by other supporters such as Metcalf.

Types of speciation other than through geographic isolation are at least as important as allopatric speciation.

There are probably several routes by which species can be formed and sympatric speciation is one likely route emphasized the importance of host shifts that isolate sympatric populations and Mayr stated that host races of phytophagous animals provide the only case where incipient sympatric speciation seemed to be possible. If host race formation can lead to speciation of plant parasites it can also be important among animal parasites. Rapid evolutionary rates and such extensive adaptive radiation as seen among parasites is not easily explained by allopatric speciation over extended time periods. The large numbers of sibling species are equally hard to explain by this model. Bush concluded that host races of plant parasites evolved sympatrically and that these were undoubtedly the progenitors of the many sibling species so often found to be sympatric on different hosts.

Bush suggested that the establishment of new host races may require only minor alterations in the genome. His basic model accounts for speciation involving only two alleles at each of two loci; one locus controlling host selection and the other controlling survival in the host. One allele at each locus carries these traits adapted to host species A and the other alleles enable the parasite to discover and exploit host species B. If host species A and B have different phenologies, and the parasites adapt to these differences, reproductive isolation is reinforced. Bush provided several examples of this allochronic isolation involving host shifts which can occur only through "a narrow window in space and time." For example, on Mount Shasta in California a host shift of the fruit fly, Rhagoletis indifferens, from bitter cherry, Prunus emarginata, to introduced domestic cherry, P. avium, could occur only at about 5,000 feet during the last two weeks of July, whereas the fly occurred from sea level to 9,000 feet and from May to October.

Hosts with different phenologies or breeding cycles may effectively isolate their parasites allochronically and may play an important part in sympatric speciation. Triggering of reproductive activity may be initiated by the host as in the rabbit and hare fleas, insects parasitic on others,and possibly among blood-feeding Mallophaga on birds.

From the gene-for-gene hypothesis Day predicted the number of host races that a set of resistant varieties will select for. With 19 genes for resistance in apple, for example, each of which may have two phenotypes, resistant or susceptible, there would be 219 or 524,288 races of a parasite adapted to exploit fully the range of apple varieties. This may seem an extreme number, but we have every reason to infer that a parasite must be closely attuned morphologically, physiologically and biochemically to the host, and, therefore, such extensive race formation may be necessary and realistic. Once the races have differentiated in this way subtle, ecological or temporal isolation could easily promote the independent differentiation of populations.

Since most parasites are small, usually with narrow tolerances to environmental factors, they are susceptible to minor spatial or temporal change. When tolerances are narrow slight differences between habitats may cause isolation where habitats may be only 100 meters apart. If reproductives or dispersing individuals live only a few days, a week's difference in phenology may prevent gene flow between populations. Many more ecotones exist for parasitic species and it is in these intermediate and changeable zones that Stebbins sees the cradle for rapid evolution. For such small, short-lived, precisely adapted organisms as parasites, evolution will operate in miniature: short times, small spaces, but impressive results.

Speciation in Parasites

Parasite speciation and host–parasite coevolution should be studied at both macroevolutionary and microevolutionary levels. Studies on a macroevolutionary scale provide an essential framework for understanding the origins of parasite lineages and the patterns of diversification. However, because coevolutionary interactions can be highly divergent across time and space, it is important to quantify and compare the phylogeographic variation in both the host and the parasite throughout their geographical range. Furthermore, to evaluate demographic parameters that are relevant to population genetics structure, such as effective population size and parasite transmission, parasite populations must be studied using neutral genetic markers. Previous emphasis on larger-scale studies means that the connection between microevolutionary and macroevolutionary events is poorly explored.

Species and Speciation

Understanding the general patterns and processes of speciation is fundamental to explaining the diversity of life. Parasites represent ideal models for studying speciation processes because they have a high potential for diversification and specialization, and some groups live in conditions that are ripe for sympatric speciation. The host represents a rapidly changing environment and a breeding site, which makes the number of diversifying factors potentially larger for parasitic than for free-living organisms. However, parasites are almost totally ignored in the general evolutionary literature on speciation processes. Even in the parasitology literature, fundamental studies of parasite speciation are scarce. Most parasite-population studies concern parasites of humans and domestic animals, and studies on natural, undisturbed populations are needed. Because ecological factors that shape microevolutionary patterns might also reinforce long-term macroevolutionary trends,we focus on parasite life histories, and parasite population dynamics and their influence on population genetics. The first step toward identifying the evolutionary processes that promote parasite speciation is to compare existing studies on parasite populations. Crucial, and novel, to this approach is consideration of the various processes that function on each parasite population level separately (from infrapopulation to metapopulation). Patterns of genetic differentiation over small spatial scales provide information about the mode of parasite dispersal and their evolutionary dynamics. Parasite population parameters inform us about the evolutionary potential of parasites, which affects macroevolutionary events. For example, small effective population size (N_e) and vertical transmission can initiate founder-event speciation, whereas natural selection can cause either adaptive or ecological speciation in parasite species with a large N_e. In 1995, Nadler considered the various microevolutionary processes that structure parasite populations and summarized all the empirical evidence available at the time. Subsequently, the wider use of molecular techniques has achieved major advances and prompted a new synthesis.

Special Features of Parasites

Parasitism is one of most successful modes of life, measured by the number of times it has evolved independently and the diversity of extant parasite species. Practically every animal species is infected by at least one parasite species and, even without strict host specificity, there are at least as many (and possibly more) parasitic than free-living species. According to Price the most extraordinary adaptive radiations have been among parasitic organisms, and he was one of the first to stress the need for studies on their evolutionary biology. Although other authors agree, relatively few have taken advantage of this purported opportunity. However, an increased interest for host–parasite cospeciation studies has paralleled the development of software programs for testing cospeciation.

Parasites represent a diverse group of biologically different organisms that are united only by their common lifestyle, which is spent either in or on the host, feeding on the tissues of the latter and causing it harm. Generally, parasites differ from free-living animals in their more-specialized feeding behaviour and intimate host associations. Although some features that are considered unique to parasites are also present in other small, specialized animals, we can identify some general characteristics:

(i)   Hermaphroditic, parthenogenetic and asexual reproduction are common;

(ii) The generation time is usually short; and

(iii) `Parasite populations are highly fragmented, with many populations also experiencing strong seasonal fluctuations in size.

These features will influence the N_e, evolutionary rate and other population genetics parameters which, in turn, affect microevolutionary and macroevolutionary processes.

Speciation in parasites can be accelerated by coevolutionary arms races and by adaptive radiations, such as after host switching. However, speciation might also be triggered by non-adaptive processes, depending on N_e. Although the overall population can be extremely large, aggregated distribution in the host population might make infrapopulations too small to minimize the role of genetic drift compared with gene flow and/or natural selection. Wahlund effects can occur (subdivided populations that contain fewer heterozygotes than predicted), whereas colonization of a new host individual and seasonal change in prevalence can cause systematic founder events. This might stimulate speciation by drift, which is described as the transilience speciation concept by Templeton, but its role in speciation is still debated.

For a neutral allele, the probability of fixation (P) in a deme increases as Ne decreases, roughly following Equation I:

$$p = 1 / 2N_e$$

But,

$$N_e = 4FM/F + M$$

(F=number of females; M=number of males).

N_e decreases as the skew in the sex ratio increases, following Equation II. Thus, if two males mate with 50 females, $N_e \approx 8$. Also, asexuality reduces N_e because of the increased linkage of the genome. N_e is expected to be decreased further by the aggregated transmission of infective stages, and when the contribution of infrapopulations is disproportionate to their effective size. Criscione and Blouin have developed a conceptual framework using a subdivided-breeders model to estimate N_e. A recent model highlights the impact of variance in the clonal reproductive success of trematode larvae and the rate of adult selfing on N_e. Previously, this first parameter has been overlooked in population genetics models and shows that, in general, parasitic life cycles are unlike those used in classical models.

Many parasite species also have a shorter generation time than their host, which results in a faster turnover (either per year or per host generation) of either neutral or slightly deleterious alleles. The average conditional fixation time (t) of a selectively neutral allele depends on N, the population size, as shown by Equation III:

$$= 4N \text{ generations.}$$

The average fixation time for a neutral mutation in an organism with a generation time of a month and a population size of 40 (and $N \approx N_e$) is ~13 years. Thus, the genetic composition can change rapidly in small populations, which enables speciation by peak shifts to occur. However, mutations that are selectively advantageous will be lost through drift, and their probability of fixation will increase as N_e increases.

The strong potential for diversification can also be ascribed partly to more frequent opportunities for sympatric speciation in parasites than in free-living organisms. These might result from narrow habitat selection either within or between host species. Furthermore, many host–parasite relationships might evolve according to co-evolutionary arms races, which fuel speciation as predicted by Thompson and Cunninghamand by the models of Kawecki. Ecological specialization can lead to speciation if certain conditions are fulfilled, and pea aphids are a well-known example of this. However, Kawecki shows that genetic trade-offs for performance on different host species are not necessary in the case of co-evolutionary interactions between a host and a specialized parasite species. Another modelling approach confirms that parasites have the characteristics required for the evolution and maintenance of adaptive diversity, with especially strong specialization and diversification expected in parasites with direct life cycles that spend several generations on the same host individual (such as pinworms and lice). Indeed, examples of ecological speciation in animal parasites such as the blood fluke *Schistosoma* are available, but controlled, comparative experiments are needed.

Parasites have a Higher Evolutionary Potential than Free-living Species

It has been demonstrated that endosymbiotic bacteria and fungi have higher rates of molecular-sequence evolution than closely related, free-living lineages. The combination of population fragmentation, extinction and recolonization patterns, regular bottlenecks, and asexual reproduction reduces the N_e and should lead to an increased rate of fixation of nearly neutral and slightly deleterious mutations. Many parasite species have these characteristics so the genetic composition can change rapidly in small populations. Several cophylogenetic studies show a higher rate of sequence evolution for the parasite than the host. In case of the chewing lice, this is linked to frequent founder events because of the small population size and vertical transmission between their pocket gopher hosts. Another correlation between an increased evolutionary rate and the transition to parasitism has been described for parasitic wasps. Again, the authors associate the parasitic lifestyle with an increased frequency of founder events and an increased selection pressure fuelled by coevolutionary arms races.

Macroevolution - The Geography of Parasite Speciation

Phylogenetic trees provide an indirect record of the speciation events that led to present-day species. By constructing species-level phylogenies and comparing the geographical distribution of sister taxa, the relative contribution of the different speciation modes can be inferred. The main problem with this approach is that the current distribution of a species is not necessarily a reliable indicator of its historical geographical range because the geographical range of a species evolves and can change considerably over short time periods (e.g. post-glacial expansion induced by the Quaternary Ice Ages). Therefore, it is possible that interspecific phylogenies cannot rigorously test alternative hypotheses concerning the geography of speciation.

Because the host constitutes the principal environment of a parasite, the geography of speciation might be inferred more readily in host–parasite systems. Tracking the evolutionary path of the host seems to be more clear-cut than tracking an entire ecosystem for a free-living species. Several statistical methods can reconstruct the ancestral host by reconciling the phylogenetic trees of host and parasite. However, results of reconciliation analyses should be interpreted with caution. Frequent host-switching can both obscure and mimic phylogenetic congruence. Absolute or relative timing of host and parasite trees can assess the possibility of synchronous speciation and, similarly, branch lengths can be taken into account in the latest version of Tree Map. Nevertheless, it is difficult to discriminate between some scenarios. For example, two parasite sister taxa found on a single host species can arise by either sympatric speciation or allopatric speciation in geographically isolated host populations followed by secondary contact. However, different solutions can be tested statistically because gene trees are amenable to mathematical modeling, and the multitude of methods available provides additional tests. Also, life-history information about the host and parasite (such as distribution range and parasite-dispersing capability) can be taken into account to choose among the optimal solutions obtained by Tree Map 2.

Generally, the speciation modes are classified according to either the geographic scale at which they occur or the population genetics events underlying them. The figure below reflects the first view, although we stress that population genetics parameters must also be included. The figure shows the influence of the different parasite-speciation modes on the phylogenetic branching pattern, and, thus, on the degree of congruence between host and parasite phylogenies. In addition to speciation by host-switching, sorting and duplication events also produce incongruent patterns. In most associations studied to date, phylogenetic congruence is either imperfect or absent; most associations represent a combination of cospeciation and host-switching. These macro-evolutionary trends are influenced by the life-history features of both host and parasite, such as host specificity and mobility. Examples of strict cospeciation occur in systems in which host-switching is prevented by the asocial lifestyle of the host and the low mobility of the parasite. Examples include the rodent (lice), and insect–symbiont associations where the bacteria that are needed for host reproduction are transmitted maternally.

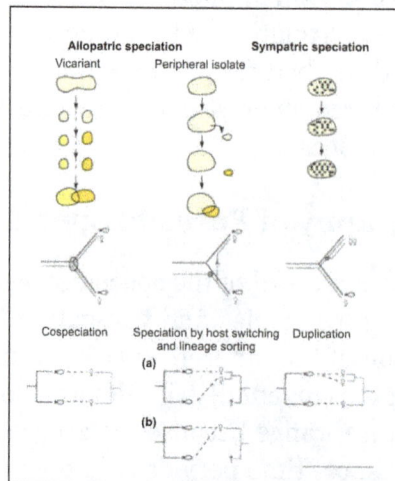

The geography of speciation and the major speciation modes and their
phylogenetic correlates applied to parasites.

In figure, An ancestral parasite species can be subdivided geographically together with its ancestral host species (vicariance). Although repeated speciations of host and parasite results in mirror-image phylogenies (vicariant allopatric cospeciation), two other possibilities exist. (a, b)Host-switching involves the movement of a small subset of a species into a new geographical area. This can be followed by speciation through a peripheral-isolates mode (a) or the new host will be added to the species range of the parasite (b). In free-living organisms, the reduction in gene flow depends on their dispersal capabilities and the magnitude of the geographical barrier, whereas, in parasites, the magnitude of gene flow depends on the transmission mode and dispersal capabilities of the parasite and the degree of sympatry between the old and the new host species. Finally, sympatric speciation occurs when species arise in the absence of a physical barrier. Brooks and McLennan define sympatric speciation as speciation on the same host species, which is equivalent to synxenic speciation defined by Combes.

Other frequently used terms in coevolutionary studies are intra-host speciation and parasite duplication. Here, we define speciation by host-switching as allopatric speciation but, depending on the parasites involved, this might also be a form of sympatric speciation when infective, free-living stages of both host-adapted populations are in syntopy. It is important to establish whether intrinsic mechanisms (evolved in the parasite) or extrinsic barriers to gene flow are involved. However, as indicated by Le Gac and Giraud, the study of speciation modes might benefit from including the mechanisms that prevent gene flow, rather than geography alone.

Cospeciation has been studied mainly at the macroevolutionary level but this process is also observed at the population level. Rannala and Michalakis studied the effect of population-level processes on patterns of cospeciation using the coalescent theory of population genetics. Demographic parameters such as N_e and transmission rate heavily influence both the occurrence and the probability of detecting cospeciation (i.e. identical host and parasite gene trees). Again, this shows that ecology has an important role in speciation processes, and stresses the need for complementary information on basic demographic parameters for both parasite and host.

Microevolution - Fragmentation of Parasite Populations

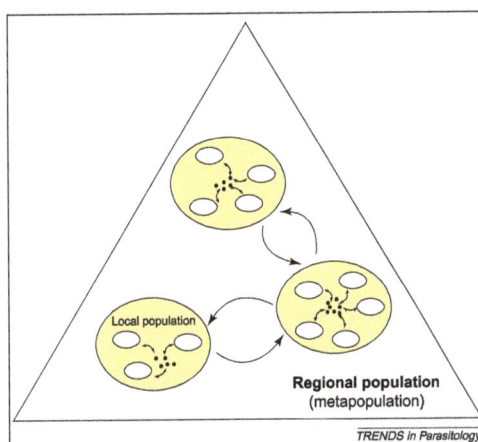

Simplified representation of parasite populations that are fragmented at the host level (infrapopulation – small white circles) and the spatial level (local population at regional scale – large yellow circles).

The fragmented nature of parasite populations can be observed on many levels. Combe distinguishes fragmentation along three scales: space; host species; and host individual. Further subdivision includes the various intermediate and paratenic host species that are used by many parasites, their aggregated distribution in the host population and the site specificity. This fragmentation greatly complicates the ecological genetics of parasites. In particular, the influence of gene flow, genetic drift and natural selection differs between levels. Therefore, we focus on the infrapopulation and metapopulation levels, predict which host and parasite traits influence genetic diversity, gene flow and population structure at each of these levels, and compare this with empirical data from the literature to review current population genetics studies in parasites. Principally, we consider spatial fragmentation of the adult parasites.

In figure, the hypothetical parasite has a direct life cycle (no intermediate hosts), with infective stages (black dots) leaving infrapopulations to be recruited in others. In intestinal helminths, for example, parasite eggs released in the faeces of one host individual will subsequently join infrapopulations in other host individuals. Note that, if all populations are connected by arrows (parasite gene flow), the population structure equals panmixia, with random mating between all individuals. In this case, the regional population (metapopulation) rather than the local population is the unit of evolution. The other extreme is completely isolated local populations, which all evolve independently.

Infrapopulation Level: Effective Population Size and Mating System

An important component of speciation theory is the deme concept because this is the unit on which either selection or drift operates. Demes are random mating populations, so when applied to parasites this denotes the adult ('sexually' reproducing) parasites that usually inhabit an individual host organism (the infrapopulation). However, the infrapopulation on one host can either comprise several demes at different sites on the host individual or it can be part of a deme when gene flow between host individuals is high. Demes are affected by inbreeding, mortality and immigration from other demes. The relative influence of natural selection and genetic drift depends on the parasite N_e, gene flow and mating system. Therefore, factors that influence inbreeding, gene flow and N_e must be evaluated to predict the deme structure.

Table: Ecological and natural-history factors that might influence the population genetics structure of parasites.

Factors that increase genetic structure	Factors that decrease genetic structure
Immobile or asocial hosts.	Highly mobile hosts (definitive, intermediate and paratenic) or vectors.
Unstable heterogeneous external environment.	Stable, homogeneous, external environment.
Complex life cycle with many specific, obligate hosts.	Persistent (long-lived) life-cycle stages in environment or definitive host.

Suitable parasite niches patchily distributed in space or time.	Uniform availability of parasite niches in space and time.
Small effective parasite population size.	Large effective population size.
Highly aggregated distribution among hosts.	Uniform distribution of parasites among hosts.
Parasite is predominantly self-fertilizing.	Parasite predominantly outcrossing.
Frequent population extinction followed by reestablishment.	Stable populations with rare extinctions.
Equilibrium between migration and natural selection.	Non-equilibrium between migration and natural selection.
Short generation time (non-coding DNA).	Long generation time.
High host specificity.	Low host specificity and/or many reservoir hosts.
Host-to-host transfer or host-mediated dispersal (vertical transmission).	High parasite mobility (horizontal transmission).

Demes, or infrapopulations, of parasites are usually short-lived and survive no longer than an individual host. New demes are formed continuously in new hosts from subsets of parasite larval stages drawn from the population gene pool. Only a few parasite taxa, in which offspring re-infect the same, long-lived, hosts as their parents over several generations, have demes that are somewhat permanent over time. Establishing the deme structure helps to predict the speciation mode. Deme size is an important factor in determining whether speciation evolves by selection or drift. For example, small, isolated demes of parasites with a direct life cycle are more likely to show population genetics patterns similar to that of their host resulting in host–parasite cospeciation at the macroevolutionary scale than demes enlarged by immigrants from other hosts.

Typically, the F_{st} index is used to describe population differentiation. However, interpreting F_{st} among parasite infrapopulations is difficult because it is influenced by factors such as drift, inbreeding and gene flow. Nevertheless, here are some examples:

(i) Infrapopulation size: The two nematodes *Ascaris suum* and *Ostertagia ostertagi* have similar life cycles, infect livestock and are obligately outcrossing. Infrapopulation sizes in *A. suum* are smaller (a few dozen per host) than in *O. ostertagi* (thousands per host) and, thus, are more susceptible to genetic drift. This results in stronger population subdivision than in *O. ostertagi*. The large N_e is assumed to be the main factor behind the unusually high within-population diversities observed for trichostrongylid nematodes.

(ii) Reproduction mode: Outcrossing trichostrongylid nematodes have large N_e, which results in high within-population diversity. By contrast, hermaphroditic juveniles of *Heterorhabditis marelatus* produce clumped patches of offspring that mate and leave the insect host. As a consequence, infrapopulations originate from a few maternal founders, which results in extremely low N_e and low within-population diversity. Parasitic nematodes of plants have a mainly parthenogenetic mode of reproduction and much lower overall mitochondrial DNA diversity than either *Ascaris suum* or trichostrongylids. In mixed-mating systems the frequency of selfing can depend on the number and size of other individuals in the infrapopulation (e.g. cestodes. Self-fertilization in *Echinococcus granulosus* results in infrapopulations with substantial heterozygote deficiencies. Asexual amplification of *Schistosoma mansoni* in intermediate hosts (snails) has little effect on the overall genetic diversity that characterizes adult infrapopulations in the vertebrate hosts (*Rattus rattus*) but slightly increases the genetic differentiation between infrapopulations at the local scale. The presence of a second intermediate host in the life cycle can serve to assemble packets of metacercariae representing many different clones, thus, ensuring genetically diverse infrapopulations of adult worms in the definitive host. By contrast, with the digenean *Fascioloides magna*, the persistence of aggregated encysted metacercariae of the same clone in the environment, is responsible for the presence of identical multilocus genotypes within hosts and a strong genetic differentiation between infrapopulations.

(iii) Premunition: In schistosomes, premunition (immunological cross-reaction stimulated by resident adult schistosomes against incoming larval parasites) was thought to restrict the genetic composition of the deme to the initial colonizers. However, concomitant immunity might operate in a genotype-specific manner, which selects for more genetic heterogeneity in male than female schistosomes, and contributes to a sex-specific genetic structure with evolutionary implications.

Metapopulation Level - Host and Parasite Mobility

At this level we assess how differences arise between parasite populations from various localities. Local adaptation is an important step towards speciation that is influenced by the interaction between gene flow and natural selection. If N_e is extremely small, genetic drift will swamp local adaptation. Simulation studies highlight the importance of host and parasite dispersal and infrapopulation processes. Higher parasite migration relative to host migration is predicted to increase the local adaptation of the parasite. Thus, gene flow and N_e are the main players at this level. Gene flow is influenced by several characteristics of the parasite, the host and the habitat, and some factors also correlate (e.g. life cycle pattern and parasite mobility). Because these factors are species-specific, patterns in different parasite groups vary accordingly.

Marine Parasites

Marine parasites may be small in size, but they can be present in very high numbers and put together can weigh even more than all the top predators in an estuary or bay ecosystem. They play an important role in keeping their host population from growing out of control—allowing them to exert power over food webs and ecosystem function. High parasite diversity is even an indicator of a healthy ecosystem.

Parasites have colonized the oceans of the globe and survive at the expense of a host that has no choice but to accept them.

Lampreys

This strange fish is considered as one of the oldest vertebrates on Earth. With inward curved teeth, lampreys cling to the skin of a fish then pierce a hole and suck the blood flowing from the wound. We find some species of lampreys that are even more voracious, going deep inside the gills of their hosts.

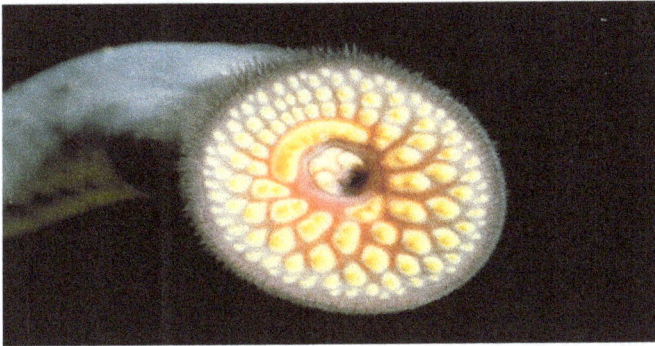

Candiru

The candiru is a well-known fish in the Amazon River. Indeed, some people in this geographical area have made a day the unfortunate experience to urinate in the water. This fish whose senses are particularly developed, followed the flow of heat and installed directly into the urethra. Apart from this accidental interference, the candiru often step into fish gills to feed on the blood and flesh of its victim.

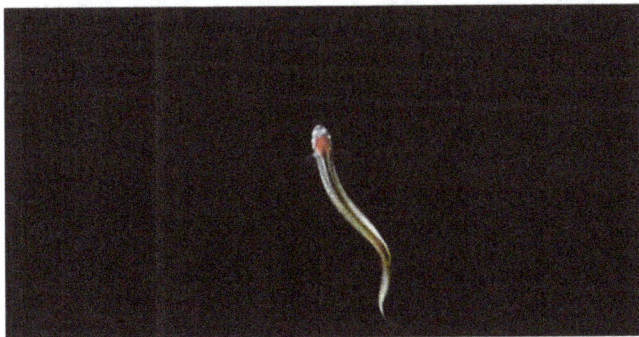

Leeches

Maybe the most famous of all parasites, over 100 marine species are currently registered. They can be found in the seven seas of the world hung on fishes, turtles or amphibians. Leeches have some amazing features: they breathe through the pores of the skin, have two hearts and they are highly resistant to pollution.

Silver Pearl Fish

The Silver pearl fish built its home in an unusual place: the anus and part of the digestive tract of a sea cucumber. This fish stays and circulates freely in this place sheltered from predators. Unfortunately for the sea cucumber, some of these fish-pearls are deleterious parasites that eat the internal organs of those which hosted them.

Cymothoa Exigua

This sea lice doesn't look like the one which attacks the hair of your children. This aquatic crustacean parasite only pink snappers. Making its way into the mouth of the host through the gill, the parasite will divert blood flow irrigating the tongue to its own advantage. Subsequently, the language will atrophy and the parasite will functionally replace the missing organ. The fish will be able to use the parasite just like a normal tongue. This example is exceptional because it is the unique case of parasitism leading to complete and functional organ replacement.

Sacculina Carcini

It is not good to be a crab in the ocean. In addition to many predators, the crab was chosen as a target by a crustacean parasite. This parasite sneaks through the joints of its claws and devours it gradually from the inside. After some time, a protuberance appears at the genital apparatus of the crab. At this stage the crab is no longer able to fix its claws, to reproduce, and it has also the duty to take care of the parasite eggs as if they were its.

Copepod

This parasite affects all groups of marine animals, from the smallest (sponge) to the largest (cetaceans). This parasite has a way of sticking on the body of its guests (in the gills, in the outer tissues, in the organs or even in the eyes). As an exemple, the copepod is fixed on the eyes of the Greenland from its birth to its death. The parasite just pumps physiological fluids through a tube making the shark blind.

Schistocephalus Solidus

This aquatic parasite lives in several hosts during its existence. We could say that this is a migratory parasit. It begins its life as an egg living in the belly of a bird. Once expelled with the feces, they are in turn consumed by a snail. This second host walks in the marshes where it's a desirable prey for the Killi (little fish). Once ingested, the parasite will infect its host and modify its behavior to get it captured more easily, for example making it resurfaces abnormally. A bird passing through will swallow the unfortunate prey and the cycle will run once again.

Gyrodactylus Salaris

This tiny parasite mostly lives on of freshwater fish skin on which it is clung thanks to sixteen sharp hooks. In order to feed, the parasite releases a liquid that dissolves the skin of the fish. Then the parasites will sucks this " soup " made of mucus and dissolved skin. At the end, lesions on the skin cause secondary infections that can lead to the extinction of many fishes. This parasite is seriously considered by the authorities.

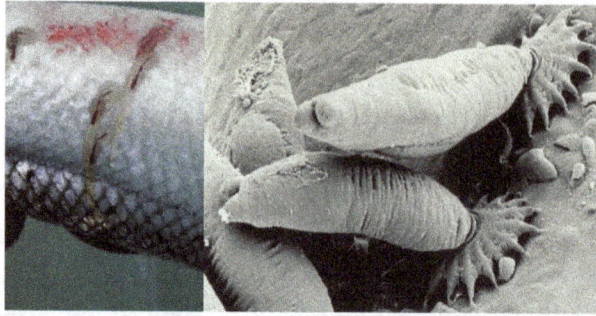

Bothriocéphale

Here's a kind of parasite we know well: the long worm. The bothriocéphale has an uncommon existence. He is born with millions of other eggs in the small intestine of a large mammal. Released in the feces of the host, it will then usually be swallowed by a crustacean, itself swallowed by a bigger predator that will be finally consumed by a mammal that can be a human being. As a piece of advice, it is recomended to cook freshwater fish flesh to avoid worms to survive. Swalloded, the worm can lead to severe complications such as anemia.

Fungal Parasites of Animals

Fungi are able to grow parasitically on a great variety of animals. In fact, it would be difficult to find an animal that doesn't have fungal parasites. When a biologist speaks of animals he or she doesn't just mean the big things like lions, tigers, dogs, cats and elephants. To a biologist an animal is any organism that is multicellular, non-photosynthetic and digests its food inside its body. This definition differentiates animals from plants and fungi. It also excludes (with some intriguing exceptions) the multitude of simple organisms that we now know to be genetically only distantly related to the three large groups.

Biologically speaking the kingdom of animals includes everything from the tiniest microscopic beings to large vertebrates, including ourselves. With such a wide variety of organisms to colonize it's not surprising that parasitic fungi come in a great variety of forms and have an equally great diversity of methods to do their work. It is also not surprising that the ones attacking worms or insects are quite different from those attacking birds or humans. Because of this division of labour among parasitic fungi we will examine them according to the group they attack.

Invertebrate Animals

An invertebrate animal is an animal without a backbone. This includes an immense diversity of organisms ranging from such simple things as corals, sponges and jellyfish to more complex animals like insects and lobsters. Of the approximately 1 million animals known to science about 950,000 (95%) are invertebrates. In fact the group is so large and complex that it would only be possible to discuss their fungal parasites in detail by subdividing it further. Instead we will examine a few examples.

Nematodes

Nematodes or roundworms comprise a large group of animals found in almost every habitat on earth. They are exceedingly numerous. Prof. E. O. Wilson has stated that if the surface of the earth were suddenly removed its shape could still be discerned by the ghostly outline of living nematodes. Four out of five species of animals are nematodes and most of these remain unnamed. With such a mass of living organisms we should not be surprised to learn that many fungi are able to use them as a source of nutrition. Several groups of fungi are specialized for detecting, infecting and consuming these little worms.

Fungi attacking nematodes are often divided into two functional groups, the endoparasites and the predators. Endoparasitic fungi develop entirely within the nematode and really have no life outside the animal while the predacious ones grow extensively in environments rich in nematodes and lay various sorts of traps to capture them.

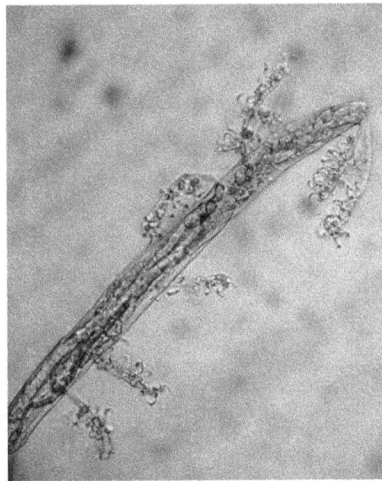

The picture at right depicts Harposporium anguillulae, a typical and common endoparasitic form. The large banana-shaped structure running diagonally from lower left to upper right is a single nematode. Numerous hyphae of H. anguillulae can be seen inside its body giving rise to spore-bearing structures that have broken through the nematode's cuticle and are producing sickle-shaped conidia (asexual spores). This nematode was orginally obtained from a sample of garden soil where it had been foraging for bacteria, fungal spores and other small bits of nutrition. In the process of this indiscriminant feeding it had ingested one or more of the sharp, curved conidia of H. Anguillulae which had become lodged in its oesophagus. The conidia had then germinated and produced the hyphae filling the nematode's body. Recent research has shown that species of

Harposporiumare in fact asexual forms produced by species of the ascomycete genus Podocrella. This is of great interest to mycologists because Podocrella species, members of the family Clavicipitaceae, are parasites of insects and other arthropods. This means that Podocrella species are among several groups of parasitic organisms that depend upon two or more hosts (prey) in their life cycles.

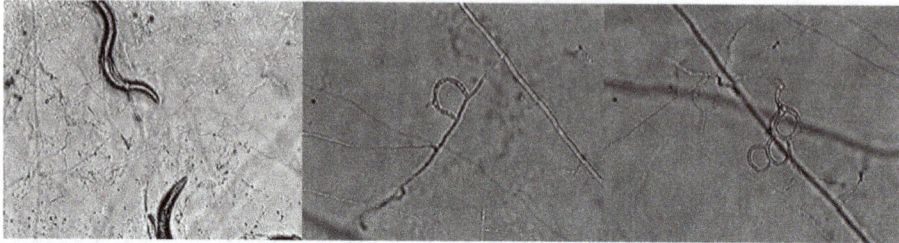

Predaceous fungi employ a variety of means to capture nematodes. These fungi grow abundantly in habitats where nematodes are abundant, producing modified sticky branches that adhere to the nematode cuticle. When a nematode comes into contact with one of these branches it gets stuck and cannot get loose. These special structures may take the form of simple short branches, knobs or loops. In some cases they have the ability to close down and form a strangle-hold on nematodes that attempts to pass through them. The picture at left shows a nematode caught by a species of the nematode-trapping fungus Orbilia. The picture above shows the traps themselves, a series of sticky loops and arches place strategically in the path of potential prey.

Insects

Insects live among fungi, pushing their way through hyphal masses and brushing against their spores. Many insects even eat fungi. With such close contact between insects and a fungus is not at all surprising that some fungi have developed the ability to live as parasites on insects. Like the fungi that attack nematodes many of those attacking insects are highly specialized and have sophisticated means of locating and infecting their prey. They may also be host-specific and occur on a few or even single species of insect. The number and diversity of fungi parasitic upon insects is so great that is is impractical to try to deal with it here.

Of the many fungi infecting insects three groups stand out as especially numerous. These are the orders Entomophthorales, Laboulbeniales and Clavicipitaceae. The picture above illustrates these three common groups. At the left is Entomophthora muscae a common parasite of flies in our area. Infected flies often land on an upright plant stem, turn bottom up and die. The parasite, which has been growing inside the insect, grows out of its recently killed prey and produces masses of white spores along each of the abdominal segments. You can easily find this fungus in the garden during late summer. Just look for grasses or other plants with an inverted dead fly clinging to ther tops. If the fungus is not obvious, place the fly on a piece of glass or plastic food wrap and enclose it so it will not dry out. By morning you will have a halo of spores surrounding the fly.

The middle picture is of Laboulbenia philonthi, a member of the Laboulbeniales. Members of this group grow attached to the outside of the insect; the black structure at the bottom is the point at which it is attached. Although external, the fungus does have an absorption cell called a haustorium inside the insect at the point of attachment. Although clearly parasitic the Laboulbeniales probably do little harm to their hosts. They are equivalent to fleas and ticks on mammals.

The fungus at right is Cordyceps variabilis. It was infecting the larva of a beetle inside a dead log, which has been cut away to show the relationship between host and parasite. Cordyceps, a member of the Clavicipitaceae, produces its spores inside the swollen head-like structure above the log. The Clavicipitaceae are fairly common in temperate parts of the world but become exceedingly abundant in the tropics.

The picture shows a dead moth colonized by the fungus Cordyceps bassiana. This one was photographed by Karen Vanderwolf in a cave in New Brunswick where it was growing at a temperature lower than 10 degrees C. This fungus is known to produce a sexual stage typical of the Clavicipitaceae. In fact, that stage was found for the first time in 2001. The asexual form, shown here and known as Beauveria bassiana, has been recognized since the early nineteenth century when the Italian mycologist Agostina Bassi demonstrated it to be the cause of 'muscardine disease' in silkworms.

Vertebrate Animals

Vertebrate animals can serve as hosts for parasitic fungi. Although vertebrates make up only a small part of the animal kingdom they display some diversity in both aquatic and terrestrial habitats. Aquatic habitats support quite different kinds of fungi than terrestrial ones so it is not surprising that fish and amphibians harbour fungal parasites having little in common with those infecting birds and mammals.

Fish

Most of the fungi parasitizing fish are not really fungi at all and instead belong to the phylum Oomycota. Although once thought to be fungi we now know that the Oomycota and more closely related to certain algae. Species of several genera of Oomycota are known to attack fish. Home aquarium owners will undoubtedly recognize the genus Saprolegnia, a notorious group of parasites that can wipe out an aquarium in a short period of time. Species of Saprolegnia get into the fish through open wounds and rapidly penetrate its tissues with broad tubular hyphae. Infected but still-living fish may be seen swimming with a tuft of hyphae trailing out behind them.

A few "true" fungi also parasitize fish, including some species of Exophiala and Fusarium. Although we usually think of Fusarium species as plant parasites they can become parasitic on a variety of animals, including humans.

Amphibians

Amphibians, including frogs, toads, newts and salamanders, may be air-breathing in their adult stages but generally are not very resistant to dryness and must remain in aquatic or very moist habitats. Because of the moisture that usually surrounds them they are susceptible to many of the same parasites as fish. Even woodland salamanders can be infected with species of Saprolegnia.

More recently biologists have begun to experience severe declines in the populations of frogs in many parts of the world. Although the cause of this decline is still not entirely understood a major factor appears to be the the fungus Batrachochytrium dendrobatidis which infects the skin of frogs causing severe leasons and finally death. The disease is called chytridiomycosis. Batrachochytrium dendrobatidis is a member of the Chytridiomycota, a group of true fungi having motile spores that can swim in search of aquatic prey. Why this fungus has become prevalent in just the last few years is not clear. Scientists have suggested global warming, pollution, habitat disturbance and other causes but the answer is still not known.

Birds and Mammals

Birds and mammals are together a smaller group than fish, and of these birds are the more numerous. Apparently swimming and flying are the way to success. These terrestrial, warm-blooded animals have a variety of fungal parasites. Of course the ones infecting humans and domestic animals are the best-known while those parasitic on wild mammals and birds remain largely unstudied.

The fungi causing mycoses or diseases in birds and mammals are often divided into several groups depending upon the site of infection. These groups are:

1. Superficial Mycoses: These are fungi that grow on the surface of the body or on hair shafts. They do not invade living tissues and usually cause no symptoms. They may often be present but undetected. One of the most common of these is Malassezia furfur, a relative of the smut fungi which may cause a mild dermatitis or discolouration of the skin. It can also be the cause of dandruff.

2. Cutaneous Mycoses: A very common group of fungi found in the outermost layers of dead skin. Although they do not invade living tissues they produce enough enzymes and other metabolites in the outer layers of skin that they elicit a response in the host that is usually seen as reddening at the site of infection. These fungi, commonly called dermatophytes, are the cause of athlete's foot, ringworm and nail disorders. Candida albicans, a member of this group that also grows on mucus membranes in the mouth, anus and vagina can cause more serious infections under certain conditions.

3. Subcutaneous Mycoses: This is a group of fungi that are probably not primarily parasitic and only become so if they enter a wound. Generally they remain localized and form small masses of hyphae in the affected tissues. They may at times invade bones. Subcutaneous mycoses may remain on the host for many years and form leaky swollen lesions.

4. Dimorphic Systemic Mycoses: Fungi in this group are called dimorphic because they can grow as hyphae or in a single-celled budding form, depending upon where they are. In a petri dish at room temperature they grow as "normal" hyphal colonies but in living tissue they begin to behave like yeasts and travel through the body as single cells. They are called systemic because they have the ability to spread to various organs. Typically the spores are inhaled with dust and germinate to form colonies in the lungs. They may remain there and cause tuberculosis-like symptoms or they may spread to other parts of the body. The diseases they cause can be serious or even fatal. Members of this group tend to be quite common in certain endemic areas and quite rare elsewhere. The picture at right is of Ajellomyces capsulatus, the fungus responsible for histoplasmosis, a disease prevalent around the southern Great Lakes and St. Lawrence Valley. On the left side of the picture are several small round objects; these are the asexual spores of the fungus and are commonly called Histoplasma capsulatum. The panel on the right side of the picture shows these spores close-up where you can plainly see the characteristic "fingers" extending out from the surface. To the right of the left panel is a mass of hyphae and bedspring-like coils. This is where the sexual ascospores are produced.

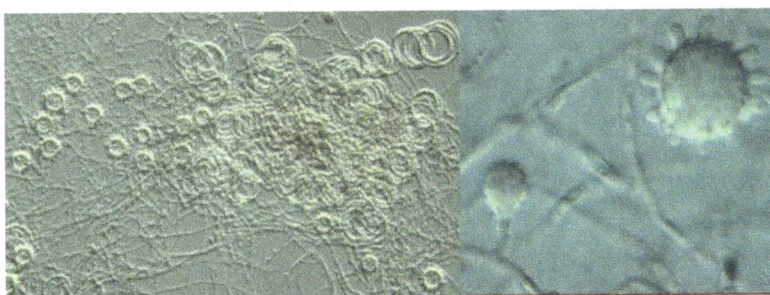

5. Opportunistic Systemic Mycoses: This group includes fungi that are able to invade tissues within the body but are not dimorphic. They are opportunistic because they normally exist as non- parasitic forms and only become virulent when the immune system has been compromised in some way. This can occur through the use of immunosuppressant drugs following organ transplant surgery or because of diseases such as AIDS and diabetes that suppress the immune system. The use of chemotheraputic agents, corticosteroids and other chemicals having a profound effect on the body may also allow them entry. These fungi are not in any kind of "balance" with the host and can invade rapidly and destructively.

Branches of Parasitology

Paleoparasitology

Paleoparasitology is the study of parasites found in archaeological material. The development of this field of research began with histological identification of helminth eggs in mummy tissues, analysis of coprolites, and recently through molecular biology.

In 1987 the first description of techniques used for the recovery of parasite eggs from archaeological materials was published. Since that time, the exploration of archaeological remains has expanded and new techniques have been devised. The most important development during the past decade has been the application of molecular biology techniques for the recovery of ancient parasite DNA. Also, chemical digestion of archaeological sediments was introduced for the effective recovery of parasite eggs from all types of archaeological deposits.

The Pioneering Period - Paleoparasitology (also archaeoparasitology) is the study of parasites in ancient material. The first report of ancient parasites is Ruffer's diagnosis of *Schistosoma haematobium* eggs in kidneys from Egyptian mummies. RUFFER used histological sectioning and staining for the identification of the eggs. Although a few other pioneering papers appeared in the first half of this century recording parasite eggs in archaeological material, the field really developed in the 1960s and was finally named in 1979.

Initially, parasitologists analyzing coprolites tried different flotation techniques. These are effective in unconsolidated sediments where parasite eggs are well preserved. However, standard clinical techniques were not effective when parasitologists turned their attentions to coprolites. Coprolites are desiccated and sometimes mineralized feces. To analyze coprolites, the trisodium phosphate rehydration technique was introduced. This technique was adapted from methods used to rehydrate desiccated zoological specimens in museums. Further experiments demonstrated that the trisodium phosphate technique was effective when applied to coprolites. These studies showed that trisodium phosphate at 0.5% concentration in aqueous solution results in the reconstitution of the eggs and larvae of parasitic worms. It was shown that the egg shells and anatomical features of the larvae such as the esophagus and intestine are visible after application of trisodium phosphate. Thus, the application of this simple technique allows for the microscopic diagnosis of parasitic worms.

In the late 1960's and early 1970's, there was a burst of paleoparasitology studies as the trisodium phosphate technique was widely applied to coprolites from Utah, Arizona, Colorado, and Nevada. These pioneering efforts can be characterized as a "discovery" phase. During this time, it was demonstrated that parasitism dates back to remote times and that prehistoric humans were hosts to a wide variety of parasites. In 1967, Aidan Cockburn pointed out that coprolite studies had a great potential for defining the evolution of infectious disease in relation to cultural evolution. He urged parasitologists to interpret their data from an epidemiological perspective. Cockburn's message reached paleoparasitologists in North and South America. Still, it took several years before his recommendations were followed and broader paleoparasitological interpretations were scarce through the 1970's and 1980's. But, by the late 1980's, paleoparasitological data began to be interpreted to a greater extent.

Paleoepidemiology

After the pioneering and discovery periods in paleoparasitology, researchers began struggling with new methodological questions. The consistent problems for paleoparasitologists are the diagnosis of the zoological origin of coprolites found in archaeological layers, and the diagnosis of the parasites themselves. A reference collection of desiccated feces of living mammals belonging to a national park in northeast Brazilian was prepared for comparison with the coprolites found at archaeological sites in the same region. The method shows good results when it can be applied. However, it is a long and tedious process to survey all animals to cover all the possibilities of fecal morphology.

Similar procedures are encountered regarding parasites. When the host is known, parasite checklists are very useful. Morphometric parameters must be examined, as proposed by experimental paleoparasitology. However, extinction and changes in the local fauna must be considered.

Also, when eggs of a parasite not previously known to exist in humans are found, careful evaluation of the infection must be done to determine if a true case of parasitism is represented. This problem was encountered with acanthocephalan eggs in coprolites from Utah and Arizona. The eggs were identified as *Moniliformis clarki*. Analysis of the dietary constituents of the coprolites and the biology of *M. clarki* showed that this was a parasite of humans. Therefore, careful paleoparasitological analysis showed that *M. clarki* was a common parasite of prehistoric Indians before indigenous dietary patterns changed. In other cases, careful analysis reveals "false parasitim" when eggs of a parasite are consumed and passed through the intestinal tract without hatching.

During these two decades paleoparasitology advances relied on morphological parameters of parasite remains. Ligth microscopy has been the main tool for scientists. However, other techniques including immunology and electron microscopy have been introduced. HORNE experimented with transmission electron microscopy (TEM). Although HORNE did not recommend that TEM replace light microscopy, TEM did allow for the identification of internal parasite egg structures. Scanning electron microscopy (SEM) is a useful diagnostic tool. In certain cases, fungal spores, and especially pollen can be confused with parasite eggs when only light microscopy is applied. SEM allows for the examination of surface features that can be used to distinguish them. Also, SEM is a very useful tool for the characterization of helmith larvae. Immunological tests have significant potential for paleoparasitology. FOUANT was the first to apply immunological analysis to parasite remains. Her application of ELISA to possible *Entamoeba histolytica* cysts proved negative. Immunofluorescence stains were successfully applied to identify *Giardia lamblia* cysts in coprolites from Kentucky. In our opinion, immunology has a great potential for identifying parasite remains.

Once the methodological problems discussed above were resolved, great advances in paleoparasitology occurred. Parasite infections were identified in various locations and a picture began to emerge of the distribution of parasites and prehistoric migration routes of their hosts. It can be said that paleoparasitology has reached a stage of metamorphosis from a descriptive stage to a period of true contributions to the pathoecology of parasitism.

Quantitative studies also showed interesting epidemiological patterns. Ancient hunter-gatherers were shown to have a reduced parasite fauna relative to agricultural populations. Also, hunter-gatherer parasitism is dominated by zoonotic species whereas agricultural populations had more human-specific parasites. Subsequent comparison of parasite prevalence in coprolites with bone lesion prevalence (porotic hyperostosis) in skeletons showed a correlation between parasitism and anemia. Comparisons of the pathoecology of prehistoric agricultural villages showed that the level of parasitism was dependent on the local ecology, sanitation patterns, and house style. Detailed studies of house type through 10,000 years of prehistory in the southwest United States have shown that pinworm prevalence is related to the style of house construction which affects air flow.

A different technique was developed in the 1980s that led to the recovery of data from soils from archaeological sites. In 1986, reinhard et al. adapted palynological techniques used for the recovery of pollen to the recovery of parasite eggs. By using treatments with hydrochloric acid, hydrofluoric acid, and zinc bromide heavy density solution, and adding *Lycopodium* tracer spores to the sediments, parasite eggs could be concentrated and quantified from any archaeological sediments. In subsequent years the technique was used on latrine soils from many archaeological sites and also on sediments from gardens. In these analyses, the widespread distribution of parasites was documented for urban sites in Israel and North America.

Molecular Biology and Ancient Parasite DNA

During the last ten years infectious diseases started to be diagnosed using technologies based on nucleic acid. Parasite diagnosis is the last field of clinical microbiology to use these techniques, and the role that they can play in epidemiology, prevention, and treatment of parasitic diseases is enormous. The probe based on nucleic acid for the detection of parasites consists of the use of a reporter molecule of DNA to detect specific sequences of parasite DNA or RNA. The parasite in the sample is lysed and the nucleic acid is released and denatured.

The polymerase chain reaction (PCR) is based on the replication *in vitro* of the double-helix molecule of DNA. It is used to amplify a segment of DNA situated between two regions of a known sequence. Two oligonucleotids are used as primers for a series of reactions catalyzed by an enzyme, DNA polymerase. PCR is the synthesis of millions of copies of a specific DNA segment.

Ancient DNA (aDNA) or ancient RNA (aRNA) are nucleic acids recovered from archaeological, paleontological or museum specimens. In a broader sense it can be applied to any nucleic acid recovered after death when the autolysis process was started.

Hybridization was the first technique used to recover DNA from archaeological material. The first molecular clone of animal DNA was prepared using the skin of an extinct zebra. PÄABO and WILSON et al. worked with human DNA of archaeological origin. However, relatively large amounts of DNA are needed for the hybridization technique.

In 1985 the polymerase chain reaction (PCR) was incorporated. It is sensitive and can be performed easily, permitting the use of small nucleic acid fragments from human, other animals, and plants.

The PCR technique was described by Kary Mullis in 1985, and Saiki et al. improved it. The technique was then adapted for archaeological material. Ancient DNA was then amplified from human bones and mummified tissues. The importance of PCR for archaeology was reviewed, showing the perspectives and limits of the new technique. Sex determination and the study of phylogenetic relationships in ancient populations were shown to be possible with this approach.

Protozoal Infection in Ancient Material

Helminths are the most common parasite finds in archaeological material. Eggs and larvae can be well preserved by desiccation, or, at times, even by mineralization. Even before the use of rehydration technique, parasite eggs were found in archaeological material.

Protozoan infections are not easy to identify in archaeological remains. *Entamoeba coli* cysts were recorded in the intestinal contents of a Peruvian mummy, and protozoan cysts were detected in human coprolites dated 1800 BP (before present). *Eimeria* cysts were found in deer coprolites dated 9000 BP, from Brazil.

Tissue protozan infections were even more difficult to diagnose. Small ceramic statues found in pre Columbian burial sites suggested cutaneous leishmaniasis lesions, and histopathological examinations of mummified bodies showed lesions identified as Chagas' disease. *Trypanosoma cruzi* was found using electron microscopy and the diagnosis was confirmed by histochemical techniques, in a Peruvian mummy.

Recently, seven Chilean mummies were found to be positive for *T. cruzi* by the PCR technique. This findind confirmed the presence of Chagas' disease since at least 4,000 years ago in the Andean region.

Ancient Molecular Biology - Experimental Research

Mucocutaneous Leishmaniasis

Many pre Columbian Andean populations used ritual burials where small clay statues (huacos) accompanied the body. Some of them show destructive nose and lip lesions, similar to those of mucocutaneous leishmaniasis.

Leishmaniasis prevalence is related to some ecological aspects of hosts, vectors and forest reservoirs. Natural infection is common among Marsupialia, Edentata, Rodentia, few carnivores, primitive Ungulata, and primates including man.

Rodents are very important for the epidemiology of mucocutaneous leishmaniasis but the infection is usually asymptomatic. However, the importance of the role of forest rodents in the maintenance and transmission cycle of the disease is not sufficiently understood.

In an attempt to study the role of rodents in the transmission cycle of leishmaniasis, PCR was applied to a sample of taxidermized rodents collected 50 years ago.

To test the technique, experimentally infected mice were taxidermized with the same technique used in the collection. The assay was conducted on 11 young mice (Balb/c) infected with *Leishmania amazonensis* (106 promastigotes/ml) in the foot. After two months, 25 mg were collected from each animal (three controls were used) and DNA was extracted. PCR followed the method of Sambrook and the QUIAGEN protocol.

The results were positive for *Leishmania* in all infected samples and negative in noninfected mice, and allowed the study in the museum collection.

Sixty thousand specimens of rodents were taxidermized and stored in the National Museum (Universidade Federal do Rio de Janeiro). The collection was the result of the plague campaign from 1941 to 1975, covering the entire country. Some rodents were captured in endemic leishmaniasis areas. When possible, the material was collected where lesions were present and from two known endemic regions. From Baturité - Ceará state, the following species were examined: Oryzomys elliurus, O. subflavus, Kerodon rupestris, Trichomys apereoides, Galea spix, and Zygodontomys pixuna. And from Ilha Grande - Rio de Janeiro state, Rattus norvegicus, O. lamia, Proechymus dimidiatus and Phyllomys sp. Taking care to avoid contamination, DNA isolation and purification were performed using the commercial QIAamp-tissue kit.

Standard procedures were followed using taqDNA polymerase and positive and negative controls were used for each PCR reaction. Hybrydization was performed to confirm the results.

Five of 39 animals were found to be positive for *Leishmania*. The results showed that the PCR technique can be applied to epidemiological studies of the past, and also to the diagnosis of leishmaniasis in mummies with suggestive lesions.

Chagas' Disease

Fifteen years ago it was observed that archaeologists were bitten by triatomines (*Triatoma brasiliensis*) when they were doing copies of rock art in the archaeological site of Pedra Furada, Piaui state, northeastern Brazil. At that time it was supposed that the ancient artists could have been infected by *Trypanosoma cruzi* 20,000 years ago. Unfortunately, no technique was available to diagnose the microorganism in the skeletons found at the site.

PCR makes this study possible today, and we began this research line in our Laboratory of Paleoparasitology.

The first step was to test experimentally desiccated infected material. *T. cruzi* infected mice were desiccated at 39 °C and PCR was used to identify *T. cruzi* DNA. The results obtained suggest that the application of this technique to *T. cruzi* detection in archaeological material was possible.

After testing the technique, we applied it to mummified tissues collected from 7 mummies from Atacama (Museo Arqueologico de San Pedro de Atacama, Chile). The first results showed positive PCR, but hybridization is in process. The second step is to test the technique in bones and tissues from the archaeological sites of Piauí state, Brazilian northeast.

Other Parasitic Infection

Research for ancient DNA in archaeological material has also been investigated for bacteria and viruses in coprolites and mummified tissues. DNA of *Shigella flexneri* was found in pre-Columbian coprolites from Chile showing potentialities for this field in paleoparasitology. Bacterial and viral DNA in archaeological material is an open field for phylogenetics and the evolution of diseases and the future is promising.

Future of Paleoparasitology

In the past, the innovation and application of new techniques led to significant contributions of paleoparasitology to the general understanding of the pathoecology of parasitism. For example, the wide application of the trisodium phosphate technique to coprolites in the past led to profound developments in the understanding of the evolution and distribution of parasitic disease. Also, the application of soil digestion to archaeological sediments led to the definition of the nature of urban parasitism. Similarly, future discoveries in pathoecology of parasitism depend on the broad application of new techniques to new materials.

Since 1994, there has been a new emphasis on mummy studies. Parasitologists have been quick to begin devising new techniques for application to mummies. The advantage of mummies is that it is possible to recover adult and larval stages of parasites, and also that parasites of somatic tissue can be recovered. Also, ectoparasites can be analyzed from mummies. Thus, it is probable that mummies will be the main focus of paleoparasitology technique development. This is already evident in the discussion presented above. Researchers in Brazil are working intensively on the development of analysis techniques for mummies.

The techniques that will be useful in mummy studies will include DNA and immunological analyses. Tests for antibodies and antigens, and searches for distinctive parasite proteins are underway or are planned. The application of immunological test has proven useful in coprolites and will undoubtedly be useful in mummies. Also, aDNA studies will be expanded. The preservation of tuberculosis aDNA sequences in Peruvian mummies, and the experimental application of this technology to leishmaniasis and trypanosomiasis indicate that application of PCR technology will result in the recovery of parasite aDNA.

We also feel that newer developments in microscopy will facilitate the study of parasites in mummies. The laser confocal microscope will prove to be useful in characterizing lesions caused by ancient parasites. The development of the environmental SEM may also prove important since this technology does not require extensive preparation of specimens by critical point drying. Therefore, more delicate samples from mummies may be used for parasite study.

Thus, the future of paleoparasitology will expand from coprolites and soils to include mummy studies. In this way, a greater diversity of species, including protozoa, helminths and arthropods will be studied and the ecology of their diseases elucidated.

Veterinary Parasitology

Veterinary parasitology is the study of animal parasites, especially relationships between parasites and animal hosts. Parasites of domestic animals, (livestock and pet animals), as well as wildlife animals are considered. Veterinary parasitologists study the genesis and development of parasitoses in animal hosts, as well as the taxonomy and systematics of parasites, including the morphology, life cycles, and living needs of parasites in the environment and in animal hosts. Using a variety of research methods, they diagnose, treat, and prevent animal parasitoses. Data obtained from parasitological research in animals helps in veterinary practice and improves animal breeding. The major goal of veterinary parasitology is to protect animals and improve their health, but because a number of animal parasites are transmitted to humans, veterinary parasitology is also important for public health.

Diagnostic Methods

Various methods are used to identify parasites in animals, using feces, blood, and tissue samples from the host animal.

Coprological

Coprological examinations involve examining the feces of animals to identify and count parasite eggs. Some common methods include fecal flotation and sedimentation to separate eggs from fecal matter. Others include the McMaster method, which uses a special two-chamber slide that allows parasite eggs to be more clearly visible and easily counted. It is most commonly used to monitor parasites in horses and other grazing and livestock animals. The Baermann method is similar but requires more specialized equipment and more time and is typically used to diagnose lungworm and threadworm.

Haematological

Haematological examinations involve examining the blood of animals to determine the presence of parasites. Blood parasites tend to inhabit the erythrocytes or white blood cells and are most likely to be detected during the acute phase of infection. Veterinary parasitologists use blood smears, which involve placing a drop of blood onto a slide and spreading it over the surface in a thin film in order to examine it under a microscope. The blood is stained with a dye in order for the cells to be easily distinguished.

Histopathological

Histopathological examinations involve examining tissue samples from animals. A small slice of the organ suspected of being infected by parasites is mounted on a slide, stained, and examined under a microscope.

Though not technically considered a histopathological technique, skin scraping – which involves taking a small sample of the epidermal cells of a dog, cat, or other household pet – is commonly used to detect the presence of mites.

Immunological

Immunological examinations, such as indirect immunofluorescence, ELISA, Immunoblotting (Western blot), and Complement fixation test are methods of identifying different kinds of parasites by detecting the presence of their antigens on or within the parasite itself. These diagnostic methods are used in conjunction with coprological examinations for more specific identification of different parasite species in fecal samples.

Molecular Biological

Molecular biological methods involve studying the DNA of the parasite in order to identify it. PCR and RFLP are used to detect and amplify parasite DNA found in the feces, blood, or tissue of the host. These techniques are very sensitive, which is useful for diagnosing parasites even when they are present in very low numbers; they are also useful for identifying parasites not only in large animal hosts but smaller insect vectors.

Divisions of Veterinary Parasitology

Veterinary Protozoology

Veterinary protozoology is focused on protozoas with veterinary relevance. Examples of protozoan parasites:

- Babesia divergens
- Balantidium coli
- Besnoitia besnoiti
- Cryptosporidium parvum
- Eimeria acervulina
- Eimeria tenella
- Giardia lamblia (also known as Giardia duodenalis)
- Hammondia hammondi
- Histomonas meleagridis
- Isospora canis
- Leishmania donovani
- Leishmania infantum
- Neospora caninum
- Toxoplasma gondii
- Trichomonas gallinae
- Tritrichomonas foetus
- Trypanosoma brucei
- Trypanosoma equiperdum
- Veterinary Helminthology

Veterinary helminthology is focused on veterinary important helminth parasites, for example:

- Ancylostoma caninum
- Ancylostoma duodenale

- Ascaris suum
- Dicrocoelium dendriticum
- Dictyocaulus viviparus
- Dipylidium caninum
- Echinococcus granulosus
- Fasciola hepatica
- Fascioloides magna
- Habronema species
- Haemonchus contortus
- Metastrongylus
- Muellerius capillaris
- Ostertagia ostertagi
- Paragonimus westermani
- Schistosoma bovis
- Strongyloides species
- Strongylus vulgaris
- Syngamus trachea (Gapeworm)
- Taenia pisiformis
- Taenia saginata
- Taenia solium
- Toxascaris leonina
- Toxocara canis
- Toxocara cati
- Trichinella spiralis
- Trichobilharzia regenti
- Trichostrongylus species
- Trichuris suis
- Trichuris vulpis

Veterinary Entomology (Arachnoentomology)

Veterinary entomology is focused on important arachnids, insects, and crustaceans. Some examples include:

- Caligus species
- Cimex colombarius
- Cimex lectularius
- Culex pipiens
- Culicoides imicola
- Demodex bovis
- Dermacentor reticulatus
- Gasterophilus intestinalis
- Haematobia irritans
- Hypoderma bovis
- Ixodes ricinus
- Knemidocoptes mutans (causing the disease scaly leg)
- Lepeophtheirus salmonis (sea louse)
- Lucilia sericata
- Musca domestica
- Nosema apis
- Notoedres cati
- Oestrus ovis
- Otodectes cynotis
- Phlebotomus species
- Psoroptes ovis
- Pulex irritans
- Rhipicephalus sanguineus
- Sarcoptes equi
- Sarcophaga carnaria

- Tabanus atratus

- Triatoma species

- Ctenocephalides canis

- Ctenocephalides felis.

Marine Parasitology

Marine Parasitology studies fish parasites in all parts of the world Oceans. One of the less studied ecosystems on earth is the deep sea. The deep-sea is characterized by an absence of sunlight, low water temperatures, high hydrostatic pressure, weak water currents, and scarcity of food. The principal factors determining the settlement of organisms are food availability and water movement. Expensive research expeditions are needed to get access to the unique research material.

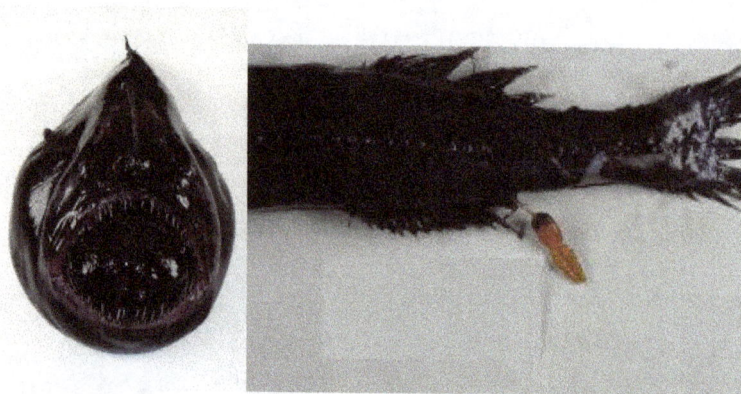

Also the deep-sea inhabits a rich parasite fauna, consisting of all major taxa. In total, 421 of the 3800-4200 known deep-sea fish species (less than 10%) have been studied for its metazoan parasites so far. These hosts harbour 621 different parasite species. It has been estimated that between 1.5 to more than 3 different metazoan parasite species can infest each single fish species also in the deep-sea environment, leading to the assumption that between 5-10,000 parasite species occur below 200 m water depth.

The pictures were taken on a research expedition to the Mid-Atlantic Ridge. Nearly nothing is known about the deep-sea fish parasite fauna in that region. However, because especially digenean trematodes and crustacean ectoparasites have been recorded to have a high biodiversity also in deep waters, these parasites were also obvious in the catche.

Medicinal Parasitology

Medical Parasitology is the branch of medical sciences dealing with organisms (parasites) which live temporarily or permanently, on or within the human body (host). There are different types of parasites and hosts. The competition for supremacy that takes place between the host and the parasite is referred to as host-parasite relationship. Accordingly, the host may have the upper hand and remains healthy or loses the competition, and a disease develops. Human parasites are either unicellular (protozoa) or multicellular (helminthes and arthropods). The parasites may live inside the host (endoparasites) or on the host surface (ectoparasites).

Endoparasites are classified into intestinal, atrial or they may inhabit body tissues causing serious health problems. Ectoparasites are arthropods that either cause diseases, or act as vectors transmitting other parasites. Human evolution and parasitic infections have run hand in hand and most parasitic diseases and methods of their transmission have been discovered thousands of years ago. Environmental changes, human behavior and population movement have a great effect on transmission, distribution, prevalence, and incidence of parasitic diseases in a community. Parasites can invade the human body in different ways; through oral route, skin, arthropod vectors or sexual contact. Host defense mechanisms consist of innate immunity which mediates initial protection against infection and adaptive immunity which is more effective. Once parasites have evaded innate host defenses, adaptive cellular and humoral immune responses are promoted against a wide array of antigenic constituents.

Diagnosis of parasitic diseases depends on several laboratory methods, imaging techniques and endoscopy in addition to clinical picture and geographic location. Parasitic diseases may be presented by a wide variety of clinical manifestations according to the tissue invaded. Direct microscopy is based on detection of the parasite by examination of different specimens (stool, urine, blood, CSF and tissue biopsies). Immunodiagnostic techniques include antigen and antibody-detection assays. Molecular-based diagnostic approaches offer great sensitivity and specificity. Recently, nanotechnology can be applied as diagnostic procedures utilizing nanodevices. Control and prevention of parasitic diseases depend on the interactions among many factors such as the environment, the human behavior, and socio-cultural factors that determine transmission and persistence of parasites.

Medical Parasitology is the science dealing with parasites that infect man, causing disease and misery in most countries of the tropics. They plague billions of people, kill millions annually, and inflict debilitating injuries such as blindness and disfiguration on additional millions. World Health Organization estimates that one person in every four harbors parasitic worms.

References

- Host-parasite-relationship, parasitology: biologydiscussion.com, Retrieved 2 February, 2019

- Types-of-hosts-in-parasitology, parasitology: biology-today.com, Retrieved 4 April, 2019

- 142-parasitic-nutrition, biology: veedhibadi.com, Retrieved 10 August, 2019

- Parasitology: cell.com, Retrieved 14 June, 2019

- The-10-most-nightmarish-marine-parasites-on-the-planet: spotmydive.com, Retrieved 18 January, 2019

- AnimalParasites: nbm-mnb.ca, Retrieved 8 March, 2019

- Elsheikha, HM; Khan, NA (editor) (2011). Essentials of Veterinary Parasitology. Caister Academic Press. ISBN 978-1-904455-79-0

- Deep-sea: marineparasitology.com, Retrieved 20 May, 2019

Chapter 2

Parasitic Protozoa

Protozoa are a type of microscopic, one-celled organisms which can be parasitic in nature. Some of the parasitic protozoa are giardia lamblia, entamoeba histolytica, balantidium coli, plasmodium malariae and plasmodium falciparum. The topics elaborated in this chapter will help in gaining a better perspective about these parasitic protozoa.

Giardia Lamblia

Giardia lamblia (syn. Giardia intestinalis, Giardia duodenalis) is a flagellated unicellular eukaryotic microorganism that commonly causes diarrheal disease throughout the world. It is the most common cause of waterborne outbreaks of diarrhea in the United States and is occasionally seen as a cause of food-borne diarrhea. In developing countries, there is a very high prevalence and incidence of infection, and data suggest that long-term growth retardation can result from chronic giardiasis. In certain areas of the world, water contaminated with G. lamblia cysts commonly causes travel-related giardiasis in tourists.

Giardia species have two major stages in the life cycle. Infection of a host is initiated when the cyst is ingested with contaminated water or, less commonly, food or through direct fecal-oral contact. The cyst is relatively inert, allowing prolonged survival in a variety of environmental conditions. After exposure to the acidic environment of the stomach, cysts excyst into trophozoites in the proximal small intestine. The trophozoite is the vegetative form and replicates in the small intestine, where it causes symptoms of diarrhea and malabsorption. After exposure to biliary fluid, some of the trophozoites form cysts in the jejunum and are passed in the feces, allowing completion of the transmission cycle by infecting a new host.

Classification and Evolution of Giardia Species

Discovery and Species Designation of Giardia

An appropriate classification for Giardia spp. is critical to an understanding of the pathogenesis and epidemiology of infection, as well as the biology of the organism. This process has been difficult for a number of reasons.

- The (presumed) asexual nature of the organism does not allow mating experiments to allow species designation. For clonal organisms in the same clade, there are no well-defined criteria for species designation; these designations remain controversial.

- Many of the earlier descriptions of Giardia spp. assumed a different species for each host and consequently overestimated the number of species. Subsequent species descriptions based on morphologic differences detected by light microscopy have probably underestimated the differences among isolates, strains, or species.

- Cross-transmission experiments of Giardia from one host to another have yielded inconsistent results.

- The available tools for distinguishing Giardia isolates have been inadequate until the recent introduction of molecular and electron micrographic techniques for classifying Giardia spp.

Giardia was initially described by van Leeuwenhoek in 1681 as he was examining his own diarrheal stools under the microscope. The organism was described in greater detail by Lambl in 1859, who thought the organism belonged to the genus Cercomonas and named it Cercomonas intestinalis. Thereafter, some have named the genus after him while others have named the species of the human form after him (i.e., G. lamblia). In 1879, Grassi named a rodent organism now known to be a Giardia species, Dimorphus muris, apparently unaware of Lambl's earlier description. In 1882 and 1883, Kunstler described an organism in tadpoles (G. agilis) that he named Giardia, the first time Giardia was used as a genus name. In 1888, Blanchard suggested the name Lamblia intestinalis, which Stiles then changed to G. duodenalis in 1902. Subsequently, Kofoid and Christiansen proposed the names G. lamblia in 1915 and G. enterica in 1920. There continued to be controversy about the number of Giardiaspecies for many years, with some investigators suggesting species names on the basis of host of origin and others focusing on morphology. For example, over 40 species names had been proposed on the basis of host of origin. Simon, on the other hand, used morphologic criteria to distinguish between G. lambliaand G. muris and accepted the name G. lamblia for the human form. In 1952, Filice published a detailed morphologic description of Giardia and proposed that three species names be used on the basis of the morphology of the median body: G. duodenalis, G. muris, and G. agilis. The species name G. lamblia became widely accepted through the 1970s. Since the 1980s, some have encouraged the use of the name G. duodenalis, and in the 1990s, the name G. intestinalis has been encouraged by other investigators. At this time there does not appear to be adequate reason to abandon the term G. lamblia, which has been widely accepted in the medical and scientific literature.

G. lamblia is pear shaped and has one or two transverse, claw-shaped median bodies; G. agilis is long and slender and has a teardrop-shaped median body; and the G. muris trophozoite is shorter and rounder and has a small, rounded median body. G. lamblia is found in humans and a variety of other mammals, G. muris is found in rodents, and G. agilis is found in amphibians.

Table: Giardia species

Species name	Hosts	Morphology by:		Molecular data
		Light microscopy	Electron microscopy	
G. agilis	Amphibians	Long and slender; teardrop-shaped median body		NA
G. muris	Rodents	Short and rounded; small rounded median body		Distant from G. lamblia
G. lamblia	Numerous mammals, including humans	Pear shaped; one or two transverse, claw-shaped median bodies		Clade with multiple genotypes
G. ardeae	Herons	Same as G. lamblia	Ventral disk and caudal flagellum similar to G. muris	Closer to G. lamblia than to G. muris

| G. psittaci | Psittacine birds | Same as G. lamblia | Incomplete ventrolateral flange, no marginal groove | NA |
| G. microti | Voles and musk-rats | Same as G. lamblia | Cysts contain two tro-phozoites with mature ventral disks | Similar to G. lamblia genotypes |

For the Giardia isolates grouped with G. lamblia on the basis of morphologic criteria discernible by light microscopy, differences that can be detected by electron microscopy have allowed the description of additional species, G. psittaci from parakeets and G. ardeae from herons. Another species, Giardia microti, has been suggested on the basis of host specificity for voles and muskrats, differences in the cyst as assessed by electron micrography, and by differences of the 18S rRNA sequences compared with G. lamblia of human origin.

Genotypes of G. Lamblia

Molecular classification tools have been of great value in understanding the pathogenesis and host range of Giardia isolates obtained from humans and a variety of other mammals. The first study of the molecular differences of G. lamblia isolates was a zymodeme analysis of five axenized isolates, three from humans, one from a guinea pig, and one from a cat, using six metabolic enzymes. Zymodeme analysis consists of the typing of organisms based on the migration of a set of enzymes on a starch gel in the presence of an electric field. The migration depends on the size, structure, and isoelectric point of these enzymes. Since these properties are a function of the primary amino acid sequence, differences in the zymodemes should reflect differences in the sequences of the genes encoding these enzymes. In 1985, restriction fragment length polymorphism analysis of 15 isolates was performed using random probes. These studies resulted in the description of three groups; group 3 was so different from groups 1 and 2 that the suggestion of a separate species designation was made. Subsequently, a number of other molecular classification studies have been performed using zymodeme and restriction fragment length polymorphism analysis. Pulsed-field gel electrophoresis (PFGE) chromosome patterns have also been studied but are of limited value for classification because of the frequent occurrence of chromosome rearrangements. Likewise, classification by surface antigens is limited by antigenic variation of the variant-specific proteins (VSPs). These studies have been very useful, but the conclusions that can be drawn from these types of data are limited by the semiquantitative nature of the data. To allow a more quantitative comparison of Giardia isolates, sequence comparisons of the small-subunit rRNA, triosephosphate isomerase (tim), and glutamate dehydrogenase (GDH) genes have been utilized in a number of subsequent studies.

These studies have all confirmed the division of G. lamblia human isolates into two major genotypes. The first consists of Nash groups 1 and 2, Mayrhofer assemblage A, groups 1 and 2, and the Polish isolates, while Nash group 3, Mayrhofer assemblage B, groups 3 and 4, and the Belgian isolates form the other major genotype. For example, the tim nucleotide sequences of the group 1 and 2 isolates diverged by 1% in the protein coding region and 2% in the flanking regions, while groups 1 and 3 diverged by 19% and the flanking regions were so dissimilar as to preclude their alignment. The small-subunit or 18S rRNA (SS rRNA) sequence shows a 1% divergence between groups 1 and 3, reflecting its more highly conserved sequence. In addition to their marked genetic differences, the two genotypes may have a number of important biologic differences. For example, the GS isolate (group 3) was significantly more pathogenic in infections of human volunteers than was the WB isolate (group 1). Group 3 organisms also appear to grow much more slowly in axenic culture than do genotype 1 organisms.

Table: Genotypes of G. lamblia

Proposed designation	Nash group	Mayrhofer assemblage	Origin	Hosts
Genotype A-1	1	A (group 1)	Poland	Human, beaver, cat, lemur, sheep, calf, dog, chinchilla, alpaca, horse, pig, cow
Genotype A-2	2	A (group 2)		Human, beaver
Genotype B	3	B (groups 3 and 4)	Belgium	Human, beaver, guinea pig, dog, monkey
		C		Dog
		D		Dog
		E (or A-livestock)		Cow, sheep, alpaca, goat, pig
		F		Cat
		G		Rat

More recently, a number of additional assemblages (genotypes) have been proposed for Giardia isolates from a variety of mammals. These isolates are morphologically identical to human G. lamblia, but sequences of their protein-coding regions differ. These studies have allowed the identification of a dog isolate that is genetically distinct from human G. lamblia. Giardia isolates from dogs have been notoriously difficult to axenize, in comparison to isolates from humans, leading to the proposal that Giardia isolates from dogs were different from cat or human isolates. However, a few dog isolates have been axenized and characterized. To further evaluate the zoonotic potential of dog Giardia, suckling-mouse infections were established from 11 consecutive infected dogs. On the basis of sequence analysis, these were assigned to two assemblages (C and D) that were quite distinct from assemblages A and B. A PCR-based study of nine fecal isolates from dogs found that one of the nine was similar to human isolates while the other eight were different. These results suggest that most dog isolates are genetically distinct from those found in humans and have little or no potential for zoonotic transmission.

Separate assemblages (E through G) have also been proposed for hoofed livestock, cat, and rat isolates. This same sequence-based study demonstrated that G. microti was a member of this compilation of seven assemblages. Further studies of Giardia obtained from cattle have demonstrated that some of the isolates belong to the livestock assemblage (assemblage E) while others belong to assemblage A (genotype 1) and thus may have the potential for human infection. Assemblages C through G have not yet been isolated from humans, suggesting the likelihood that some genotypes of G. lamblia have a broad range of host specificity that includes humans while others appear to be more restricted in their host range and may not pose a risk of zoonotic transmission. Whether these seven assemblages should be considered separate species should await further data and consensus.

Host-Parasite Coevolution

Whenever members of a grouping of parasitic organisms parasitize different hosts, it is appropriate to address the question of coevolution of the host and parasite. For Giardia spp., the ability to do so has been limited because of the uncertainty about the host specificity of a specific Giardia species or genotype as well as the problem of appropriate classification of isolates. The development of good molecular classification tools and a better understanding of host specificity make it reasonable to address the possibility of coevolution of different species or genotypes of Giardia with their hosts. The greater difference between G. lamblia and G. ardeae than among the G. lamblia genotypes supports the idea that the divergence between G. lamblia and G. ardeae accompanied the divergence between birds and mammals. However, G. lamblia and G. ardeae are closer to each other than to G. muris, the opposite of the expected finding, since mice diverged from other mammals more recently than from birds. Sequence information is not yet available from G. agilis to allow a similar comparison of the amphibian and mammalian Giardia species. The closer-than-expected relationship between G. ardeae and G. lambliacould perhaps be explained if transmission of Giardia between birds and mammals was followed by divergence of the two lineages with their hosts.

Giardia and other Diplomonads as Early-Branching Eukaryotes

Traditionally, all living organisms have been classified as prokaryotes or eukaryotes, and some still argue for retaining the two major divisions. However, the most widely accepted classification now utilizes three major divisions, Archaea (archaebacteria), Bacteria (eubacteria), and Eukarya (eukaryotes), which can then be divided into kingdoms. With either classification system, G. lamblia is clearly a eukaryotic organism and has been considered a member of the protozoa, the more "animal-like" of the unicellular eukaryotes. These protozoan organisms have traditionally been classified by their morphology into flagellates, ciliates, amebae (rhizopods), and sporozoa. Thus, G. lamblia was classified with the flagellated protozoans, including the kinetoplastids (e.g., Leishmania spp. and Trypanosoma spp.), parabasalids (e.g., Trichomonas vaginalis), and Dientamoeba (e.g., Dientamoeba fragilis). Giardiahas been placed in the order Diplomonadida (two karyomastigonts, each with four flagella, two nuclei, no mitochondria, and no Golgi complex; cysts are present, and it can be free-living or parasitic) and the family Hexamitidae (six or eight flagella, two nuclei, bilaterally symmetrical, and sometimes axostyles and median or parabasal bodies), along with the mole parasite Sppironucleus muris and the free-living organism Hexamita inflata. Some of the higher orders of classification do not appear to be phylogenetically valid, such as the placement of all flagellated protozoans together. However, the family Hexamitidae does appear to be a monophyletic group.

One recent classification system with one prokaryotic and five eukaryotic kingdoms has retained Protozoa as one of the six kingdoms, but most recent proposals have suggested abandoning the term "protozoa" in favor of the more general but at the same time more precise term "Protista".

Recent classifications of the eukaryotic microbial organisms have depended primarily on molecular comparisons. An ideal molecular classification system would be based on a gene(s) that is required for all life and that is sufficiently highly conserved across all forms of life that accurate alignments allow comparison and classification of all organisms. In many ways, rRNA has been the most useful gene for molecular comparisons, because rRNA sequences are highly conserved across life and because the function of the rRNA is so central to the biology of the organism. Therefore, the most widely accepted classification scheme has been based on SS (18S) rRNA sequences. Based on comparisons of SS rRNA sequences, G. lamblia was proposed as one of the most primitive eukaryotic organisms (308), along with T. vaginalis and the microsporidia. The use of SS rRNA sequences to place Giardia as an early-branching eukaryote has been criticized because of the high G+C content of the SS rRNA of Giardia (75%) and because Giardiaspp. are parasitic organisms; artifacts may be introduced by high rates of mutation accompanying host adaptation. An analysis of the early-branching eukaryotes suggested that the basal position of Giardia was an artifactual result of long-branch attraction due to a greater evolutionary rate of Giardia. Analysis of the large subunit of RNA polymerase II and reanalysis of the eF1 and eF2 sequences also supported the idea of the long-branch artifact. However, no such effect was shown for the eRF3 tree. In addition, the absence in G. lamblia of the highly conserved N-terminal domain of eRF3 found in other eukaryotes including T. vaginalis suggested the divergence of G. lamblia before the acquisition of the N-terminal domain.

It should also be noted that the phylogenetic placement of Hexamita, a free-living organism in the same family (as determined by morphologic criteria), avoids artifacts due to high G+C content or parasitism. The G+C content of the Hexamita inflata SS rRNA is 51%, and H. inflata was found to be monophyletic with Giardia. The classification of H. inflata with G. lamblia is also supported by a comparison of the glyceraldehyde-3-phosphate dehydrogenase sequences of G. lamblia, Trepomonas agilis, H. inflata, and Spironucleus sp. A comparison of the SS rRNA sequences of Giardia, Hexamita, Trepomonas, and Spironucleus, all diplomonads, showed that all were phylogenetically related and that the last three comprised one clade while Giardia occupied another clade. These sequence-based comparisons have led to the following proposed classification system for Giardia: kingdom Protozoa, subkingdom Archezoa (includes the phyla Metamonada and Microsporidia), subphylum Eopharyngia, class Trepomonadea, subclass Diplozoa, and order Giardiida (includes the families Octomitidae and Giardiidae). In this classification system, the diplomonads are considered to comprise the subclass, Diplozoa, and Hexamita, Trepomonas, and Spironucleus are all members of the other diplozoan order, Distomatida.

An examination of the different genes used for the phylogenic classification of Giardia shows that the genes used in transcription and translation generally yield a basal position for Giardia, although artifact due to long-branch attraction has not been ruled out. Interestingly, a phylogenetic tree of one of the genes thought to be of mitochondrial origin (cpn60) also suggests that Giardia is an early-branching eukaryote. Of the other groups of genes, the metabolic genes do not yield a clear position, but some suggest a eubacterial origin for Giardia. Cytoskeletal genes give mixed results, with actin giving an early divergence but tubulin yielding a divergence that is later than that of Entamoeba histolytica. Among other classes of genes, HSP70 and cathepsin B phylogenies suggest an early divergence for Giardia. Thus, although the answers are not conclusive, most of the data suggest that G. lamblia and the other diplomonads are among the most basal of the extant eukaryotes.

Table: Molecular phylogeny of G. lamblia

Gene	Phylogenetic position
SS (18S) rRNA	At base of tree with T. vaginalis and microsporidia or long-branch artifact
Translational apparatus	
RNA polymerase III	Later; similar to T. brucei
EF1-alpha	Early divergence; diplomonads form a single clade or long-branch attraction artifact
EF2	Early divergence or long-branch attraction artifact
Eukaryotic release factor 1 (eRF1)	Basal
Eukaryotic release factor 3 (eRF3)	Early divergence

RNA polymerase II large subunit (RPB1)	Position cannot be determined because of long-branch attraction artifact
Transcription apparatus	
Fibrillarin	Basal
Mitochondrial genes	
cpn60	Basal along with T. vaginalis and E. histolytica
Valyl-tRNA synthetase	Secondary lack of mitochondria; related to gamma-proteobacteria; branches with Arabidopsis
Cytoskeleton	
Annexins	Alpha-giardins as primitive annexins
Actin	Early divergence
β-Tubulin	Later divergence than E. histolytica
Metabolic pathways	
Triosephosphate isomerase	Eubacterial (alpha-proteobacteria) origin, unweighted parsimony puts Giardia at base
Glutamate dehydrogenase (NADP dependent)	Possibly of eubacterial origin
Adenylate kinase	Impossible to resolve order of emergence
Glyceraldehyde-3-phosphate dehydrogenase	Monophyletic with other diplomonads; later divergence
Pyruvatephosphate dikinase	No evidence for early divergence
Malate dehydrogenase	Eukaryotic, related to T. vaginalis
Malate dehydrogenase, decarboxylating	Position cannot be adequately resolved
Fructose-1,6-bisphosphate aldolase (class II)	Similar to eubacteria
Acetyl-CoA synthetase	No evidence for early divergence
Other genes	
Cathepsin B (cysteine protease)	Early divergence
Signal recognition peptide	Eukaryotic with some bacterial and archael features
HSP70 (cytosolic and GRP78/BiP)	Giardia at base, before T. vaginalis; similarities to eubacteria
CDC2 (member of cyclin-dependent kinase family)	Early branching, but after E. histolytica and T. vaginalis

Trophozoite Structure

The *G. lamblia* trophozoites are pear-shaped and are approximately 12 to 15 μm long and 5 to 9 μm wide. The cytoskeleton includes a median body, four pairs of flagella (anterior, posterior, caudal, and ventral), and a ventral disk. Trophozoites have two nuclei without nucleoli that are located anteriorly and are symmetric with respect to the long axis. Lysosomal vacuoles, as well as ribosomal and glycogen granules, are found in the cytoplasm. Golgi complexes become visible in encysting trophozoites but have not been confirmed to be present in vegetative trophozoites. However, stacked membranes suggestive of Golgi complexes have been demonstrated.

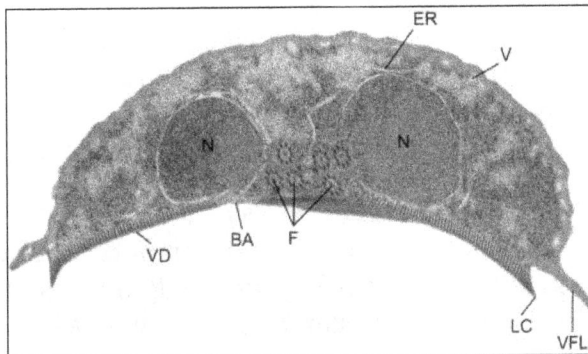

Trophozoite coronal section.

In figure, a coronal view of a trophozoite demonstrates the nuclei (N), endoplasmic reticulum (ER), flagella (F), and vacuoles (V). A mechanical suction is formed when the ventral disk (VD) attaches to an intestinal or glass surface. Components of the ventral disk include the bare area (BA), lateral crest (LC), and ventrolateral flange (VLF).

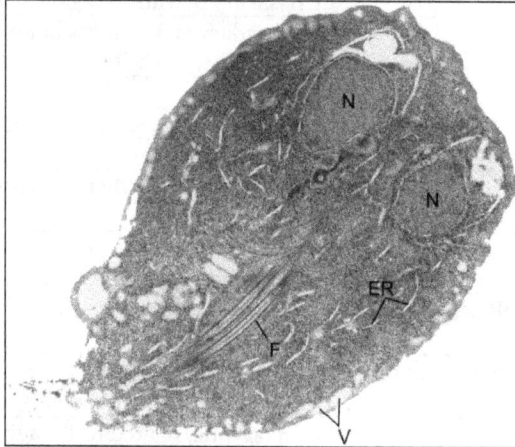

Trophozoite cross section. A cross-sectional view of a trophozoite demonstrates the nuclei (N), flagella (F), vacuoles (V), and endoplasmic reticulum (ER).

Entamoeba Histolytica

Entamoeba histolytica is an invasive, pathogenic protozoan, causing amoebiasis, and an important cause of diarrhea in developing countries. Our understanding of its epidemiology has dramatically changed since this amoeba was distinguished from another morphologically similar one, Entamoeba dispar, a non-pathogenic and commensal parasite. These two species can now be distinguished mainly through molecular and immunological procedures.

The life cycle of the parasite is represented by two forms: the cyst and the trophozoite. The cyst is the infective and non-motile form of the parasite. It is excreted in the feces and can survive for weeks in the environment. Mature cysts possess 4 nuclei and average 20 µm in diameter. The trophozoite is the motile form, with a size ranging from 10 to 60 µm. It colonizes the intestinal tract leading mainly to tissue destruction and secretory bloody diarrhea.

Amoebiasis is basically an acute disease acquired by: (i) ingestion of cysts present in contaminated food, water, or plants, (ii) through person to person contact, (iii) exposure in endemic areas, and (iv) swimming in contaminated water. Clinical manifestations range from the asymptomatic carrier state to dysenteric symptoms represented by abdominal pain and bloody diarrhea.

The organism can be prevalent in cold regions as well as tropical and subtropical regions that have contaminated water. In fact, E. histolytica is an important cause of morbidity and/or mortality wherever sewage facilities are inadequate. As is the case for other intestinal protozoan pathogens, wastewater treatment techniques are reported not to be very efficient for E. histolytica elimination possibly because of their resistance to disinfectants and the small size of the cysts. Stabilization ponds have been reported to be more effective than activated sludge for their abatement. Sedimentation and filtration can enhance the removal of cysts from wastewater.

Entamoeba

World Health Organization reported that Entamoeba histolytica affects approximately 500 million people worldwide, resulting in symptomatic diseases in 50 million and mortality in 100,000 persons. About 80-90% of infections are asymptomatic and are likely due to the nonpathogenic species E. dispar or E. moshkovskii, therefore the worldwide incidence of E. histolytica is, more likely, estimated to be 5 million cases annually, with global mortality still at 100.000 persons per annum. Very young persons, pregnant women, recipients of corticosteroids, and malnourished individuals appear more susceptible than others to amebic colitis. Entamoeba species are single cell organisms with two life cycle stages. Cysts are directly excreted in the stool and spread through the environment via contaminated water, soil, and fresh vegetables as well as unsanitary household conditions. Species cannot be differentiated based on cyst or trophozoite morphology. Following ingestion, cysts transform into vegetative forms or trophozoites, the motile stage that moves with the aid of pseudopodia and colonize the intestinal mucosa of the large bowel. Damage to the colon is caused by neutrophils that respond to infection from E. histolytica. Trophozoites can also invade the intestinal mucosal barrier and, via the bloodstream, disseminate to the liver, lung, and other sites with resultant pathologic manifestations. Drug treatments are available. Lagoons and constructed wetlands, sedimentation, filtration, flocculation, chemical and ultraviolet disinfection have all been employed for removal of cysts from water with varying degrees of success.

Asymptomatic or Non-invasive Disease

Asymptomatic infection is defined as the presence of *E. histolytica* in stool in the absence of colitis or extra-intestinal infection.

Most infections do not exhibit overt clinical manifestations and are self-resolved in few months. But, when they occur, for almost one disease case per four asymptomatic intestinally infected individuals, they are insidious and intermittent, commencing as abdominal discomfort and tenderness, bloating, irregular bowel habits, dysentery with or without blood/mucous, tenesmus with bloody mucoid diarrhea without fecal leukocytes, constitutional symptoms, toxic megacolon and finally symptoms and signs of peritonitis secondary to perforation. Moreover, the non-invasive disease could persist or progress to an invasive one in which trophozoites penetrates the intestinal mucosa.

Invasive Disease

In the invasive disease, trophozoites kill epithelial cells and invade the colonic epithelium leading to a gradual onset of abdominal pain and tenderness, weight loss, and diarrhea with bloody stools.

Amebiasis is the fourth greatest cause of mortality and the third greatest cause of morbidity by protozoa worldwide. The clinical syndrome associated with this disease is an amebic colitis resulting in a flask shaped ulcer. During this invasive stage, trophozoites, with ingested erythrocytes, replicate at a high rate in the host tissues. However, in the tissue lesions, cysts production decrease and are never found.

Unusual manifestations of *amebic colitis* include acute fulminant or necrotizing colitis with abdominal pain and distension accompanied by rebound tenderness, which usually requires surgical intervention to prevent mortality. *E. histolytica* infection can also occasionally result in the formation of an amebic granuloma or ameboma, which is an inflammatory thickening and granulation of the intestinal wall around the ulcer mimicking colonic cancer in appearance.

Extra-Intestinal Disease

E. histolytica is a facultative pathogen, exhibiting a wide range of virulence including the ability to metastisize to other organs, leading to extra-intestinal amebiasis. The liver is the most commonly affected organ through access via the portal venous system. Because it receives the bulk of the venous drainage from the right colon, the right lobe of the liver is 4 times more likely to be involved than the left lobe. Complications of amebic liver abscess include its rupture and secondary bacterial infection. The collection of pus in the liver as an abscess may present acutely, with fever and right upper abdominal tenderness, and subacutely, also with abdominal pain and prominent weight loss. Moreover, the skin and the white of the eyes become yellow or jaundiced. It is assumed that it is closely related to a history of alcohol abuse.

Hematogenous spread of trophozoites to other sites, such as lungs or brain is rare but does occur. The second most common extra-intestinal site, after the liver, is the lungs. Pulmonary infections follow direct extension of the hepatic lesion across the diaphragm and into the pleura. Amoebic brain abscess results when E. histolytica trophozoites invade the central nervous system via the bloodstream, and is frequently lethal. In Tunisia, diagnosed 3 cases of amoebic cerebral abscess: 2 men aged respectively 33 and 43 years and one woman aged 56 years who were successfully operated on. Among these three important manifestations, another rare form is cutaneous amebiasis, arising as a complication of amoebic dysentery, usually deriving from contaminated skin or a continuous contact with exudates containing virulent trophozoites.

Taxonomic Classification of the Agents

Taxonomy

Entamoeba histolytica was first described as *Amoeba coli* by Lösch in 1875 after the stool diagnosis of a mortal dysentery case and took its present name in 1903.

It is classified in the domain of *Eukarya*, kingdom of *Protista*, phylum of *Sarcomastigophora*, and the subphylum of *Sarcodina*, which contains both free-living and parasitic members. *Sarcodines* of the class *Rhizopodea* and the subclass *Lobosea*, undergo locomotion and feeding using pseudopods. This protozoan belongs to the order *Amoebida*, the family *Endamoebidae* and the genus *Entamoeba*. Among the eight species of human intestinal amoeba (*E. histolytica, E. dispar, E. moshkovskii, E. hartmanii, E. Bangladeshi, E. polecki* also called *E. chattoni, E. hartmanni* and *E. gingivalis*), *Entamoeba histolytica* is the only pathogen and the others are considered non-pathogenic and rarely cause disease in humans. However, showed that *E. moshkovskii* induced diarrhea, bloody stool and weight loss in susceptible mice and diarrhea in Bangladeshi children. This is emphasizing the pathogenicity of this protozoan. In consequence, two *Entamoeba* species could be considered pathogenic: *E. histolytica* and *E. moshkovskii*.

On the other side, *E. histolytica* is morphologically identical to *E. moshkovskii* and *E. dispar* in both the appearance of the cysts and trophozoites stages. These similarities were initially recognized by Brumpt, but he was ignored until 1978 when tremendous advances in immunology, biochemistry and genetics proved that *E. histolytica* consisted of two species: an invasive pathogen (*E. histolytica*) and non-invasive species (*E. dispar*).

The acquired virulence of *E. histolytica* is based on its ability to secrete enzymes and proteases, enabling it to penetrate the colon mucosa through the invasion of epithelial cells and subsequently interfering with the host's humoral immune response. This pathogenic amoeba is a single cell organism that can replicate in the large intestinal tract by binary fission division. Phylogenetically, *E. histolytica* is located on one of the lowermost branches of the eukaryotic tree. Thus, it is different from higher eukaryotes by its inability of de novo purine synthesis and its lack of glutathione. It also uses pyrophosphate instead of ATP at several steps of glycolysis and has a unique alcohol-aldehyde reductase. However, it shares, with higher branching eukaryotes, mitochondrial genes that are enclosed within a biochemically inert remnant organelle. Additionally, comparing to the others eukaryotes, *E. histolytica* possesses in the RNA polymerase II promoter a GAAC element, which is a novel conserved sequence specific to the rate and the size of mRNA transcription.

E. histolytica has 8,197 genes, a genome of 20.6 MB size containing six isoenzymes, helping it mainly in the lysis of tissues and the digestion of food material and intra-luminal cellular debris.

Physical Description of the Agent

Among the species infecting humans and colonizing the intestine, three (*E. histolytica*, *E. dispar* and *E. moshkovskii*) are morphologically identical in both their cysts and trophozoites. Concerning *E. bangladeshii*, its similitude with the three aforementioned, was only reported for the trophozoite form.

E. histolytica has a typical fecal-oral life cycle composed of dormant quadric nucleate cysts eliminated in feces and invasive trophozoites that replicate in the intestine.

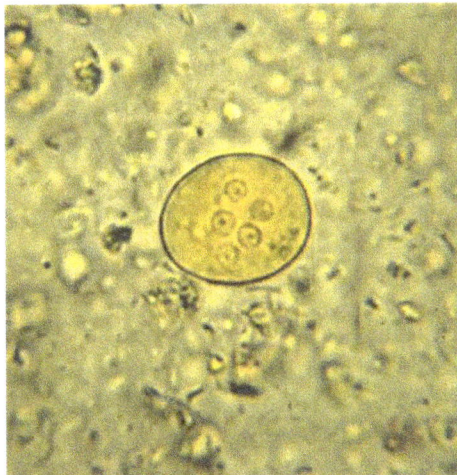

Cyst of *E. histolytica/ E. dispar/ E. moshkovskii* in raw water samples stained with iodine.

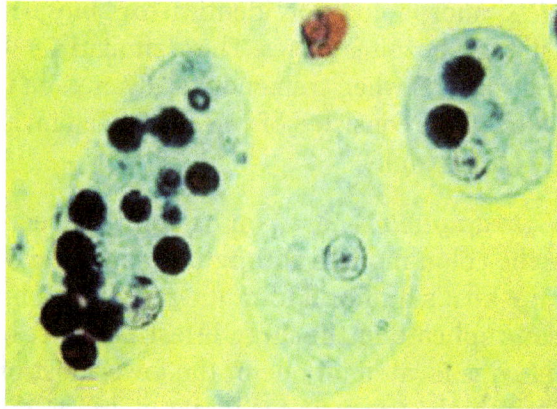

Trichrome stain of *E. histolytica* trophozoites in amebiasis.

The cyst is the infective, non-motile form which undergoes excystation during its progression in the gastrointestinal tract. Mature cysts possess 4 nuclei, have an average diameter of 20 μm, a chitin composed wall that appears smooth and refractile when viewed by light microscopy and which contains glycogen deposited in a vacuole and ribosomes that are often aggregated to form elongated bars. These cysts are directly excreted in the stool.

Vegetative forms or trophozoites are the motile stage that moves with the aid of pseudopodia, formed by the flow of ectoplasm followed by endoplasm and allowing them to move at a speed of 5 mm/s. Pseudopodia tend to be the broad blunt type, called lobopodia (Subclass of Lobosea). Trophozoites are characterized by an amorphous shape with an average diameter ranging between 15 and 30 μm, a rough endoplasmic reticulum or Golgi bodies, helices arranged ribosomes, and an absence of classical mitochondria.

Upon ingestion, these motile forms can colonize the intestinal mucosa of the large bowel.

Indeed, this vegetative stage could be represented by two forms:

   (i) The minuta or precystic stage and

   (ii) The histolytica trophozoite or magna stage as illustrated in figure.

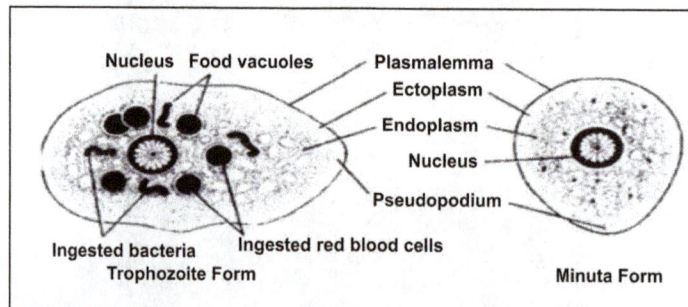

The two types of *E. histolytica* vegetative form.

  a. The minuta form (10 à 15 μm) is an intermediate stage between the trophozoite and the mature cyst and is characterized by a clear ectoplasm, a granulate endoplasm containing phagocyted bacteria and a central karyosome. It is not pathogenic, and normally lives in the lumen of the intestine and is rarely found in tissues. It undergoes encystation and contributes to the transmission from one host to another. This form is mainly present in asymptomatic and convalescent patients.

b. The histolytica form is bigger (25 à 40 μm) and contains in its endoplasm, in addition to the spherical and central nucleus, food vacuoles, phagocytized bacteria, and red blood cells. It inhabits the anterior part of large intestine.

These forms are present in at least two species: *E. dispar* and *E. histolytica*, which are morphologically indistinguishable unless the latter is observed with ingested red blood cells (erythrophagocystosis).

Life Cycle

Entamoeba histolytica is monogenetic or monoxenous, meaning that its life cycle is completed only in one host, which is usually humans, and does not need an intermediate host. However, non-human primates, some other animals like monkeys have been infected with *E. histolytica* .

The ingestion of infective mature cysts (possessing 4 nuclei), from contaminated food, water, or hands, is followed by their excystation only when they reach the terminal part of the small intestine (Ileum). The gastric juice has no effect and cannot dissolve the chitinous cyst wall. This ingestion is unfortunately, of a daily occurrence in most developing countries.

Excystation is the process of transformation of the cysts into trophozoites by disruption of the cyst wall, through the trypsin enzyme action, allowing the quadric-nucleated amoeba to emerge and to start proliferation. The amoeba undergoes a round of nuclear division followed by 3 successive rounds of cytokinesis (cell division) to produce 8 small uni-nucleated trophozoites, called amebula. Under hostile conditions, they could also produce cysts, once they aggregate to the intestine mucin layer and by accumulating a considerable amount of food in the form of glycogen, black rod-like chromatid granules, and by forming a thin, rounded and resistant cyst wall. Thus, encystation occurs in the intestine lumen, and cyst formation is complete when four nuclei are present. The signals leading to encystation or excystation are poorly understood, but research on the reptilian parasite *Entamoeba invadens* suggested that encystation might be trigged by ligation of a surface galactose-binding lectin to the surface of the parasite. After that, both stages are passed in the feces: trophozoites are usually present in loose stool, whereas cysts are found in firm stools. Because of the protection conferred by their walls, cysts are able to resist environmental conditions for days to weeks and can be responsible for disease transmission. However, once outside the body, trophozoites passed in the stool are rapidly damaged, and if ingested do not survive exposure to the gastric environment. The foregoing describes the typical asymptomatic colonization, accounting for almost 80 to 90% of all infections.

In 10 to 20% of infections, when the trophozoites colonize the intestine by adhering to colonic mucin glycoprotein via a galactose and N-acetyl D-galactosamine (Gal/galNac)-specific lectin, they feed on bacteria and cellular debris and undergo repeated rounds of binary division. Damage to the colon is caused by neutrophils that respond to the invasion. The trophozoites may also invade the intestinal mucosal barrier and gain access to the bloodstream, whereby they can disseminate to the liver, lung, and other sites with resultant pathologic manifestations. From these sites they cannot leave the host.

The figure below shows the *E. histolytica* life cycle in humans after ingestion of cysts for both kinds of infections: the asymptomatic (non invasive) accounting for almost 80 to 90% of cases and the invasive one accounting for 10 to 20% of cases.

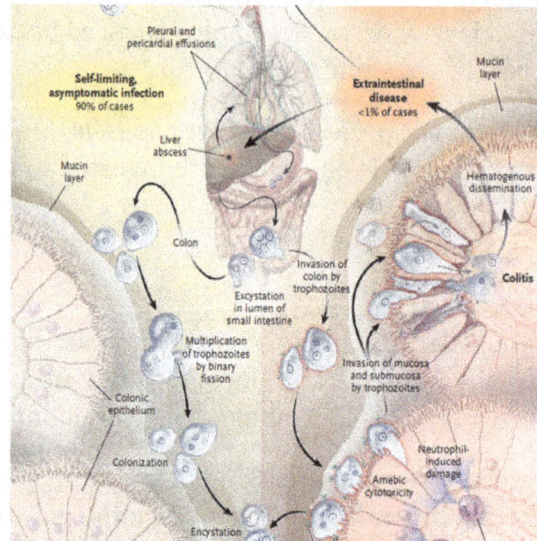

E. histolytica life cycle

Transmission

Routes of Transmission

Knowing the routes of transmission provides comprehension of how the disease is disseminated. Interventions aimed at the prevention and the control of these routes can reduce the incidence, the prevalence of diarrheal illness, and the economic impacts.

Amebiasis manifestations range from mild to severe or toxic and the course of the infection may vary in different communities based on disparities in development, standards of living, sanitation, use of stagnant water, place of residence, intense fly breeding, number of rooms and bedrooms per house, being infected by other parasites, the nutritional status, and age of infected persons. Individuals in houses with less than five rooms had amebiasis with greater infection risk compared to those living in houses with more rooms. Prevalence of *E. histolytica* infection could also be higher among members of family eating with their fingers or at the same plate.

In South Africa, a cross sectional study of stool and seroprevalence in children in a community with treated water supply and a rural site with minimal sanitation detected an important difference between them. Counter-intuitively, the children living in the first community harbored both parasites (*E. histolytica* and *E. dispar*) whereas those in the rural community harbored only *E. dispar* with a high rate, more than 50%. Jackson et al. suggested that: (i) host genetics, (ii) strain differences, and (iii) competition between the two parasites could potentially account for this striking difference.

Basically, amebiasis is acquired, like most enteric protozoa, by the ingestion of cysts present in contaminated water or food. They can survive for weeks in inadequately treated sewage/sewage products. Livestock and domestic pet handling are potential sources.

The weather can impact prevalence of infections. Extreme events such as excessive rainfall, floods, and droughts impact their abundance in the environment by the increase of the runoff or the flush of animal manure and wildlife or pets droppings into surface or ground water. Extreme weather events can also decrease water treatment effectiveness.

The foodborne and waterborne transmission remain the primary sources of infection despite ongoing investment in better sanitation infrastructure, water quality and environmental protection legislation. Food can become contaminated at several stages: during production, harvesting, handling, transport, and preparation processes. The risk of foodborne transmission is increased when food is consumed raw or undercooked. Food handlers are another possible source of infection, as are mechanical vectors such as flies and cockroaches. Contamination could also occur from cross-contamination with soiled implements, animal manure or through irrigation with contaminated water or sewage and with the use of biosolids as fertilizers or soil conditioners.

To reduce the negative public health effects of amebiasis, proposed education of the general population to stop eating raw or undercooked foods, to wash fruits and vegetables, and to wash hands before eating, as well as enforcement of environmental laws. In Brazil, estimated that the risk of acquiring amebiasis was 1.6 times greater among persons who ate raw vegetables than among those who do not eat raw vegetables. They also highlighted the causes of contamination and related it to the quality of the water used to wash vegetables and to habits of food preparation.

Entamoeba histolytica can also be transmitted through many water sources including drinking and recreational waters, lakes, and streams. The impact depends on the geographic location which is different throughout the world as well as within regions depending on the socio-economic status. There is much great reliance on water from lakes and streams for drinking water in Asia and Africa whereas they are used to a greater extent for recreation in Europe.

The spread of contamination through water could be mainly attributed to:

- A process failure within water utilities,

- The sewage contamination, or

- The presence of basic sewerage systems or latrines.

Wastewater and municipal sanitary workers have a higher incidence of amebiasis than the general population. An 11% occurrence of the complex *E. histolytica/E. dispar* in French sewer workers, and of *E. histolytica* inform of a greater number of Vietnamese farmers engaged in wastewater-irrigated agriculture or aquaculture with highest rates in people having 2-4 hours of daily contact with wastewater versus those with less exposure.

In an epidemiological study among the Oran Asli communities in Malaysia, *E. histolytica* was associated with main risk factors:

- Drinking untreated water,

- Bathing and washing in a river,

- Not washing hands after playing with soil or gardening,

- Defecation in river or bush,

- Low household income.

E. dispar, in addition to these risk factors, was also associated with:

- Age: the children less than 15 years old where the most contaminated,

- Outdoor sewage disposal,

- Consuming raw vegetables,

- The low level of mother education.

Although there were no factors associated with *E. moshkovskii* infection, others authors associated its presence with sanity conditions and lifestyles.

Reservoirs

Amebiasis is primarily a disease of humans and non-human primates, without a defined zoonotic reservoir. In non-human primates, ingested cysts proliferate in the small intestine and are carried to the colon, where the amoeba attacks the epithelial lining. Infections are usually acquired from a chronically ill or symptomatic cyst excreter.

Humans are the only known reservoirs of *Entamoeba histolytica*. Infection of other mammals (apes, cats, dogs, pigs) is infrequent, as these hosts act as neither reservoirs nor sources of infection, meaning that they do not excrete cysts with their feces and therefore they can't survive outside of their bodies. No vectors are reported transmitting amebiasis.

Incubation Period

The incubation period is usually 14 to 28 days, but could range from few days to months or years.

Approximately 4 to 10% of carriers infected with *E. histolytica* develop clinical disease within a year. Amebic liver abscess can occur years after the exposure and may follow the onset of immune-suppression.

Balantidium Coli

From the genus *Balantidium*, *Balantidium coli* is a large ciliated protozoan parasite. *Balantidium coli* have been known about for over a century, but the process of infection has yet to be discovered. It is the only known ciliated parasite to infect humans. It is responsible for the disease Balantidiasis. *Balantidium coli* is found worldwide but predominately found in the areas where pigs are raised. Other potential reservoirs include ones that hold rodents and non-human primates. More human infections happen in areas where pigs are raised. Humans are usually infected when they ingest contaminated water and food. Fecal to oral transmission is most common. *Balantidium coli* occur as a trophozoite and a cyst in the colon. The prevalence of this infection is only 1% and have been noted in Latin America, Bolivia, Southeast Asia, the Philippines and New Guinea. The prevalence of *Balantidium coli* in pigs is reported to be from about 20% to 100%.

When studying *Balantidium coli*, stool samples are taken because *Balantidium coli* does not stain well on permanent stained smears. Once out of the colon though, *Balantidium coli* must be studied immediately because it is rapidly destroyed. *Balantidium coli* can also be examined by getting a tissue sample from the periphery of ulcers during endoscopy examination (examining the inside of the body using a lighted, flexible instrument called an endoscope). Because *Balantidium coli* can occur as a cyst, cyst can also be used to study the large ciliated protozoan parasite, but this process is rare.

Genome Structure

The *Balantidium coli* genome structure continues to undergo research. Only parts ribosomal parts have been completely sequenced. At the Department of Integrative Biology, College of Biological Science, University of Guelph, N1G 2W1 Guelph, Ontario, Canada there is a unpublished journal on by Struder-Kypke, Wright AD, Foissner W, Chatzinotas A, and Lynn DH called "Balantidium coli small subunit ribosomal RNA gene, complete sequence." The small ribosomal RNA gene that was sequenced had bases 1 to 1640. Their research concluded that "Balantidium currently classified in the Vestibuliferida, did not group with the other vestibuliferids, suggesting that this order may be paraphyletic, relating to a taxonomic group that includes some but not all of the descendants of a common ancestor.

Cell Structure and Metabolism

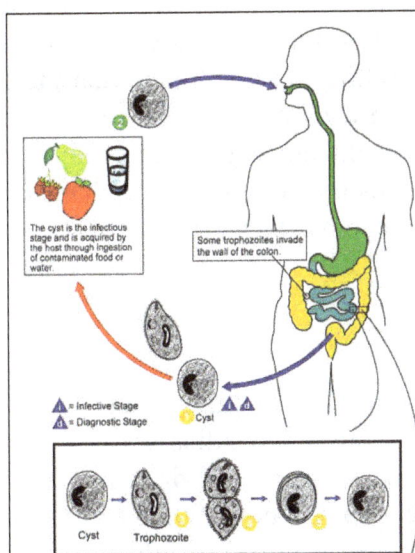

Balantidium coli are an intestinal protozoan parasite. It is the protozoan parasite of humans. It is a species of ciliate protozoan. Ciliates have 2 nuclei, macronuclei and micronuclei. The macronucleus is a long, kidney-shaped structure while the micronucleus is spherical. The micronucleus is usually next to the macronucleus. *Balantidium coli* are the only known ciliated parasite that is infectious to humans. *Balantidium coli* have two contractile vacuoles. *Balantidium coli* have two developmental stages called the trophozoite stage (reproductive stage) and the cyst stage (infectious stage).

In the trophozoite stage, *Balantidium coli* can measure between 50-130 µm long by 20-70 µm wide. When observing *Balantidium coli* unstained, it has a short ciliary covering and has spiraling motility. The two nuclei of *Balantidium coli* are clearly visible in this stage when the specimens are stained. The peristome, which is an opening at the anterior end of cell, is also visible. The peristome leads to the cytostome (cell mouth).*Balantidium coli* reproduces either by asexual transverse binary fission or sexual conjugation. In asexual transverse binary fission the bacteria grows in volume until it divides in half to make two identical daughter cells. This is how most bacteria typically grow. In sexual conjugation, a transfer of genetic material between bacteria through direct cell-to-cell contact happens. During sexual conjugation in bacteria is not an equal exchange of genetic material. There is usually a donor and a recipient. The motile trophozoite lives in the lumen of the large intestine. It feeds on intestinal bacterial flora and nutrients. The trophozoite can invade into the mucosa of the large intestine by penetrating through it. The trophozoites are released in the host's feces and this is when the cyst stage (infectious stage) begins.

The cyst stage begins right when the trophozoite is released in feces. Encystation begins. Encytation is the process of becoming enclosed in a cyst. This process takes place in the rectum. This is done by *Balantidium coli* as a survival technique. After the release of trophozoites in the feces the feces starts to dehydrates right away. Cysts can be spherical of ellipsoid and range from 50-70 µm long, 40-60 µm across, which makes it smaller than trophozoites. When a stool sample if first observed, the cysts may still have cilia, but after encystation the cilia disappear. The macronucleus and the micronucleus can still be seen in stained specimens. Because cysts exist in fecal matter, this is when contamination comes into play. Feces can contaminate water or food. Once the contaminated food or water has entered the host's digestive system, cysts can start infecting. Cysts can have cyst walls made of one or two layers, which make them tough and heavy. This allows them to pass through the digestive system without being destroyed. "There a tough cyst wall allows the cysts to resist degradation in the acidic environment of the stomach and the basic environment of the small intestine until it reaches the large intestine." Once in the large intestine excystation occurs. This is when trophozoites are produced from cysts. Cysts multiply by transverse binary fission.

Ecology

Balantidium coli are passed on through contaminated food and water. *Balantidium coli* are transmitted through a fecal-oral route. Balantidiasis (disease caused by *Balantidium coli*) can be prevented when simple and not so simple environmental changes. Of course each case of Balantidiasis different steps needs to be taken. Since *Balantidium coli* is transmitted be a fecal-oral route proper disposal of fecal material should be the first environmental change. This is simple in developed countries and not so simple in undeveloped countries. Water sanitation keeps *Balantidium coli* from contaminating water. A simple task of boiling water before consumptions may work. Another option is to get water from a source that is not near where pigs or human sewage are. Proper sewage and trash disposal away from any source of any water supply is crucial.

Pigs are the main in animal reservoirs are the main source of *Balantidium coli* cysts. There should be barriers between human and pigs. This can be done by simple fencing between human habitats and pig habitats. Pigs carry a lot of *Balantidium coli* so different water supplies should be offered to them, so that there is no contamination of pig water and human water. Personal hygiene is also important. When handling pigs hands should be washed. Hands should be washed after using the restroom, before cooking and before eating. Asymptomatic carriers should be giving antibiotics since their fecal mater can release infectious cysts. This reduces the transmission of *Balantidium coli*. The communities that are usually affected by *Balantidium coli* are poorly educated, have poor living conditions and are not developed.

Pathology

Balantidium coli cause a disease called Balantidiasis. Balantidiasis is an uncommon infection. Balantidiasis is caused through contamination and transmitted through fecal-oral route. *Balantidium coli* primarily is found in the lumen of the large intestines. It can penetrate into the mucosa layer of the large intestine and start to cause ulcers. *Balantidium coli* produce an enzyme called hyaluronidase to penetrate into the mucosa layer. Pigs are a great host for *Balantidium coli*. They are asymptomatic carriers (which mean they show no signs of carrying the infection). Pig's feces carrying vast volumes of *Balantidium coli* contaminating water sources is a huge problem. Humans who work with pigs are also exposed to *Balantidium coli*.

There is an incubation period of days to weeks after ingestion of *Balantidium coli* before an infection occurs. Most people who are infected with *Balantidium coli* are asymptomatic like pigs. Some patients can have cysts in their feces and not even know it. Common symptoms of Balantidiasis include chronic diarrhea, occasional dysentery (diarrhea with passage of blood or mucus), nausea, foul breath, colitis (inflammation of the colon), abdominal pain, weight loss, deep intestinal ulcerations, and possibly perforation of the intestine. Fulminating acute Balantidiasis is when the disease comes very suddenly and with great intensity. Hemorrhaging can occur, which can lead to shock and death. Untreated fulminating acute Balantidiasis is reported to have a fatality rate of 30%.

A stool sample is collected and a wet mount in prepared when diagnosing for Balantidiasis. The wet mount is examined for cysts or trophozoites, through a microscope. *Balantidium coli* only passes periodically through stool samples, so stool samples should be taken frequently to make a definite diagnosis. Trophozoites can also be detected in tissue. Using a sigmoidoscope, (a thin, hollow instrument used during a sigmoidoscopy procedure) the physician inspects the rectum and the sigmoid colon. They look for bleeding, ulcers, and inflammation to diagnose the cause of the diarrhea or other complaints. They also can take a tissue biopsy for inspection. When diagnosing Balantidiasis because it symptoms are similar to those of amebiasis. To distinguish the difference a careful microscopic examination needs to make.

Even though most Balantidiasis patients are asymptomatic, they should still be treated so that further transmission will not occur. There are normally three antibiotics that are given to patients diagnosed with Balantidiasis: tetracyclines, metroidozole, and iodoquinol. Tetracyclines are recommended most at 500mg four times a day for ten days. Metroidozole is prescribed at 750mg three times a day for five days. If Iodoquinol is prescribed, it is prescribed at 640mg three times a day for twenty days. If Balantidiasis is not treated the persistent diarrhea leads to high fluid loss and dehydration. If abdominal bleeding occurs, it can lead to death.

Plasmodium Malariae

Plasmodium malariae is a malaria parasite that causes a disease that has been recognized since the Greek and Roman civilizations over 2,000 years ago. Quartan, tertian, and semitertian patterns of fever in patients were described by the early Greeks. After the discovery by Alphonse Laveran in 1880 that the causative agent for malaria was a parasite, detailed studies on these organisms commenced. The early detailed work of Golgi in 1886 demonstrated that in some patients there was a relationship between the 72-hour life cycle of development of the parasites and a similar periodicity of the paroxysm (chill and fever pattern in the patient), whereas in other patients there were 48-hour cycles of development. He came to the conclusion that there must be more than one species of malaria parasite responsible for these different patterns of cyclical infection.

Eventually, the different parasites were separated and given the names that they carry at the present time. In 1890, Grassi and Feletti reviewed the available information and named *P. malariae* and *P. vivax* with the following statement: "C'est pour cela que nous distinguons, dans le genre Haemamoeba, trois espèces (*H. malariae* de la fièvre quarte, *H. vivax* de la fièvre tierce et. *H. praecox* de la fièvre quotidienne avec coutres intermittences etc.).

Plasmodium malariae has developmental cycles in the mosquito and in the primate host. When gametocytes are ingested during mosquito feeding, a process called exflagellation of the microgametocyte occurs, resulting in the formation of up to eight mobile microgametes. Following fertilization of the macrogamete, a mobile ookinete is formed, which penetrates the peritropic membrane surrounding the blood meal and travels to the outer wall of the midgut of the *Anopheles* mosquito. There, under the basal membrane, the oocyst develops. After a period of 2 to 3 weeks, depending on the temperature, many hundreds to a few thousand sporozoites are produced within each oocyst. The oocyst ruptures and the sporozoites are released into the hemocoel of the mosquito. The sporozoites are carried by the circulation of the hemolymph to the salivary glands, where they become concentrated in the acinal cells. During feeding, a small number of sporozoites (<100) are introduced into the salivary duct and injected into the venules of the bitten human, to initiate the cycle in the liver.

In the human, following introduction into the bloodstream, the sporozoites rapidly invade the liver within an hour, where, within a parenchymal cell, the parasite matures in approximately 15 days. Eventually many thousands of merozoites are produced in each schizont. Upon release, these merozoites invade erythrocytes and initiate the erythrocytic cycle. There is no evidence of quiescent liver stage forms (hypnozoites) such as are found in *P. vivax* and *P. ovale* infections in humans. However, not all liver stage forms will mature on the same day; biopsies indicate that these forms may rupture and release parasites over a number of days. Following a developmental cycle in the erythrocyte that lasts, on average, for 72 h, from 6 to 14 (average, 8) merozoites are released to reinvade other erythrocytes. Some of the merozoites develop into the two forms of gametocytes (micro- and macrogametocytes). When they are taken into the mosquito during feeding, the cycle is repeated.

Human Host

Prepatent Period

There are only a limited number of reports on the transmission of *P. malariae* to humans to determine prepatent periods. The prepatent period is defined as the time until the first day that parasites are detected on a thick blood film. Shute and Maryon reported the shortest prepatent period of 16 days for a West African strain. Boyd and Stratman-Thomas reported prepatent periods of 27, 32, and 37 days for two different strains, and Mer transmitted a Palestinian strain to three patients, in whom the periods were 26, 28, and 31 days. Prepatent periods of 23 and 26 days were reported by de Buck for four patients infected with a Vienna strain, and Boyd and Stratman-Thomas reported 28- and 40-day prepatent periods. Marotta and Sandicchi reported incubation periods (days until symptoms first appeared) of 23 and 29 days in two patients. Boyd reported on three different strains for which prepatent periods ranged from 28 to 37 days. Siddons reported a prepatent period of 30 days, and Young and Burgess reported prepatent periods of 29 and 59 days. Mackerras and Ercole reported a 24-day period for a Melanesian strain, and Kitchen reported a mean prepatent period of 32.2 days (range, 27 to 37 days) for American strains of *P. malariae*. Young and Burgess transmitted the USPHS strain of *P. malariae* to patients, and the prepatent periods were 33 and 36 days. Ciuca et al. reported prepatent periods for the Romanian VS strain ranging from 18 to 25 days. Lupascu et al. reported incubation periods of 18 to 19 days for the VS strain; in four additional patients the prepatent period ranged from 21 to 30 days. In transmission studies with a Nigerian strain involving four volunteers, the prepatent periods ranged from 24 to 33 days. Thus, as these data show, there is a wide range in the length in the prepatent period in mosquito-transmitted *P. malariae* (16 to 59 days).

Fever

The most detailed study of the paroxysm of *P. malariae* is probably that by Young et al. in which they examined 420 paroxysms. The average fever peak was 104.1 °F (rectally), with the highest recorded being 106.4 °F. Fevers (≥101 °F) ranged in duration from 5 to 32 h, with an average of 10 h 58 min. Some fevers were introduced by chills, while others were not.

A retrospective examination of induced infections with *P. malariae* was made by McKenzie et al. These data were extracted from the records of patients who were given malaria therapy for the treatment of neurosyphilis between 1940 and 1963. Prior to the introduction of penicillin for the treatment of syphilis, malaria was one of the most effective treatments for the disease. It was estimated that perhaps 20% of patients in U.S. mental hospitals had neurosyphilis, and infection with *P. vivax* or *P. malariae* was standard practice in the treatment of the disease. *Plasmodium falciparum* was less commonly used because of the difficulty of controlling infections with this species of parasite. It was believed that a combination of repeated episodes of high-intensity fever combined with a nonspecific stimulation of the immune system induced by the malaria parasite combined to destroy the spirochete. Because most African American patients were resistant to infection with *P. vivax* (due to the Duffy negative blood grouping), they were most often treated with *P. malariae*.

For these patients, the median number of days of fever of ≥101 °F was 21.9 and the median number of days of fever of ≥104 °F was 10.2. The median maximum fever for the 69 patients was 105.6 °F. One patient (S-1112) failed to exhibit fever of ≥101°F in spite of a maximum parasite count of 4,100/μl. Fever often occurred on an every-third-day pattern, as shown in figure. It is also apparent that the fever occurred just prior to an increased parasite count associated with release of a new population of parasites. Because fever regularly occurs again on the fourth day in many patients, *P. malariae* infections are often referred to as being "quartan" malaria.

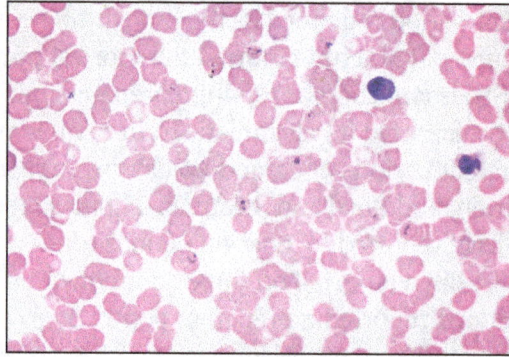

Parasitemia

Maximum parasite counts are usually low compared to those in patients infected with *P. falciparum* or *P. vivax*. This is due to several factors:

- The lower number of merozoites produced per erythrocytic cycle,

- The extended 72-hour developmental cycle versus the 48-hour cycle of *P. vivax* and *P. falciparum*,

- The preference of the parasite to developed in older erythrocytes,

- The combination of these factors that allows for the earlier development of immunity by the human host.

In the 69 patients, the maximum parasite count ranged from 1,648/µl to 49,680/µl, with a median count of 8,875/µl (10,000/µl = 0.25% of erythrocytes infected). Some patients had long periods of parasitemia and extended periods when parasite counts were >1,000/µl. These patients averaged 50.5 days with parasite counts of >1,000/µl. When the parasite counts for these patients were averaged for the first 40 days of patent parasitemia, it was apparent that the parasite count peaked at approximately 2 weeks and then remained relatively stable. The median parasite count actually did not begin to decline until 60 days or more of patent parasitemia.

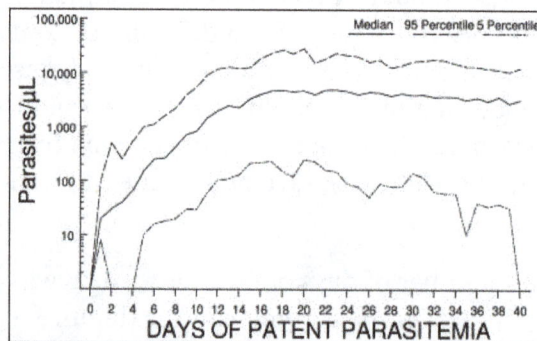

Median parasite counts during the first 40 days of patent parasitemia for 69 patients infected with *Plasmodium malariae*. Maximum parasite counts are limited in infections with *P. malariae* due to the low number of merozoites produced, 72-hour developmental cycle, and preference for older erythrocytes.

Other patients in the data studied by McKenzie et al. had been infected with *P. malariae* following previous infection with other species of malaria parasites. Forty-six patients were infected following infection with *P. falciparum*. The maximum parasite counts ranged from 312/µl to 29,825/µl, with the median being 6,608/µl. The length of parasitemia was shorter, and there were fewer days with parasite counts of >1,000/µl. The ratio between the numbers of days of fever of ≥101 °F to ≥104 °F was almost identical to that for patients with no previous infection. In addition, 39 patients had been infected with *P. malariae* following infection with *P. vivax*. The maximum parasite counts ranged from 424/µl to 19,624/µl, with a median of 9,250/µl. Only eight patients were infected with *P. malariae* following infection with *P. ovale*. The median maximum parasite count was 13,575/µl.

Recrudescence

Plasmodium malariae does not relapse from persistent liver stage parasites. However, the blood stage parasites persist for extremely long periods, often; it is believed, for the life of the human host. There have been many reports of people who have left zones of endemicity and, either following donation of blood in which the recipient developed an infection or under stress, whose infections have recrudesced after many years of dormancy. For example, Collins et al. reported on a transfusion case in which the donor had probably acquired infection with *P. malariae* in China 50 years previously. Vinetz et al. report a case of an infection acquired in Greece over 40 years (and possibly up to 70 years) previous to splenectomy and subsequent diagnosis. Because almost all of these long-term infections have been detected following transfusion donations, it is believed that the parasites have persisted in the blood at very low densities.

Preerythrocytic Stages

The preerythrocytic tissue stages develop in the liver following the introduction of sporozoites. The time required for maturation and release of merozoites from the mature schizonts to invasion of erythrocytes is approximately 15 days. The tissue stages of *P. malariae* were first described by Bray in liver biopsy specimens from sporozoite-inoculated chimpanzees. The host cell nucleus was enlarged and pushed to one side. In over 50% of the parasitized parenchymal cells, two or more nuclei were present. He was able to describe the 8-, 9-, 10-, 11-, 12-, and 12.5-day-old forms. The nuclei were always randomly distributed; there were no pseudocytomeres, no evidence of septum formation or plasmotomy, and no mature schizonts at these time points.

Lupascu et al. obtained biopsy material from a chimpanzee at 12, 13, 14, and 15 days after introduction of sporozoites of *P. malariae*. The schizonts were considered mature at 15 days. The main characteristics were enlargement of the host cell nucleus, many peripheral and internal vacuoles, no cytomeres, large clefts, red-staining strands, and plaques in the mature schizonts.

Millet et al. reported the development of preerythrocytic stages of *P. malariae* in cultures of hepatocytes from chimpanzees and *Aotus lemurinus griseimembra* monkeys. Schizonts were observed in chimpanzee hepatocytes at 8, 11, and 13 days after inoculation of sporozoites. Only one schizont was seen in *Aotus* hepatocytes at day 13.

Mosquito Host

Many different vectors have been shown to be capable, at least experimentally, of infection with this parasite. Those that have been proven to be capable of transmitting *P. malariae* to humans experimentally are also indicated. The development of *P. malariae* in mosquitoes has been described by a number of workers; the first definitive studies were carried out by Shute and Maryon on its development in *Anopheles atroparvus* mosquitoes. In the studies of Collins et al. with *Anopheles freeborni*, when incubated at a temperature of 25 °C, sporozoites were present in the salivary glands in 17 days. At day 6, the mean oocyst diameter was 12 µm, with a range of 9 to 14 µm. The oocysts continued to grow so that by day 14 they ranged from 20 to 65 µm, with a mean of 38 µm. Early differentiation and formation of sporozoites were apparent by day 14.

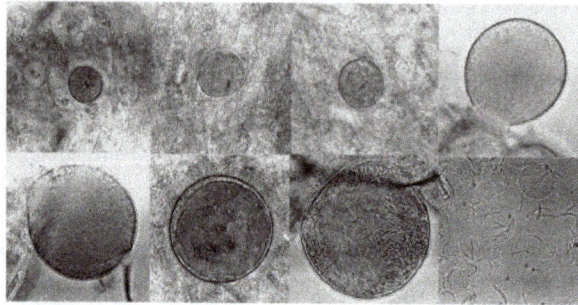

Development of oocysts of *Plasmodium malariae* in *Anopheles freeborni* mosquitoes. Top row, 10-, 11-, 12-, and 13-day oocysts; bottom row, 14-, 15-, and 17-day oocysts and sporozoites.

Plasmodium Falciparum

Plasmodium falciparum is a protozoan parasite that causes an infectious disease known as malaria. *P. falciparum* is the most severe strain of the malaria species correlated with almost every malarial death. The other 3 species that cause malaria include: *P. vivax, P. ovale,* and *P. malariae.* Humans become infected by a female Anopheles mosquito which, transfers a parasitic vector through its saliva into the blood stream. The parasite then infects the liver and undergoes asexual reproduction followed by insertion into red blood cells where an additional round of replication takes place. *P. falciparum* changes the surface of an infected red blood cell causing it to adhere to blood vessels, cytoadherence, as well as to other red blood cells. In severe cases this leads to obstructions of microcirculation resulting in dysfunction of many organs. Symptoms depend on severity of infection and can present a range of signs such as flulike symptoms, vomiting diarrhea, shock, kidney failure, coma, and death. *Plasmodium falciparum* mostly infects children under the age of 5 as well as pregnant women.

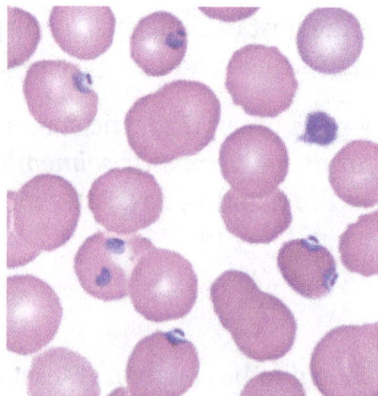

Trophozoites of P. falciparum in thin blood smears.

Pathogenesis

Transmission

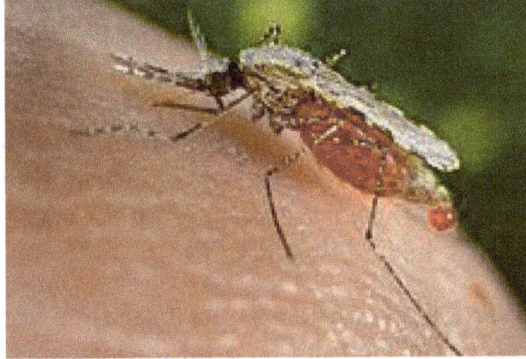

An Anopheles stephensi mosquito.

Transmission of *P. falciparum* occurs between humans and Anopheles mosquitoes. Mosquito vectors pass malaria from host to host. The parasite can infect the mosquitoes through the in take of human blood or a human may be infected by the mosquito's injection of saliva. Once the mosquito becomes infected with *Plasmodium falciparum* it transfers the disease to each new host it penetrates. Humans can rarely transfer the parasite between each other. There have been rare cases of contaminated transfused blood infecting the recipient, but seldom does this occur because of screening that takes place pre-blood donation. Mothers can also pass *P. falciparum* to their child during birth, this is also a seldom occurrence.

Infectious Dose, Incubation and Colonization

Symptoms of Malaria typically begin 8-25 days following infection however, in a few cases it can take up to a year. The late onset of incubation is due to taking an inadequate amount of anti-malaria medication. The infectious dose is not precisely known, but it is understood to be a very low number. Malaria can be observed months to years after first set of symptoms are observed. This is due to the parasites ability to lie dormant in liver cells until the environment is right for a relapse. This is mainly seen in *P.vivax* and *P. ovale*, rather than *P. falciparum*. The parasite colonizes in the liver and is then released into the blood stream and enters erythrocytes.

Epidemiology

The key to Malaria-endemic is Anopheles the mosquito's ability to live in a certain area. Temperature is also important having to stay above 20 degrees Celsius. The main areas of *P. falciparum* are South America, Africa, India, and few parts of Indonesia. The ideal location for transmission is along the equator in a warmer region. Transmission will not occur in high altitudes, colder seasons, and deserts. Malaria is considered to have arisen since the beginning of mankind, but was first discovered in blood in 1880 and found to be transmitted by mosquitoes in 1889. There are four common species of Malaria of which P. falciparum is the most severe. *Plasmodium falciparum* continues to increase in drug-resistant populations and insecticide-resistant mosquitoes leading to the prediction that the disease will only worsen over time.

Virulence Factors

PfEMP1, *P. falciparum* erythrocye membrane protein 1, is an adhesive ligand protein which is created inside of a *P. falciparum* infected erythrocyte and presented on the surface. PfEMP1 is known as a knob and is encoded by the multigene segment, Var. The protein is responsible for sequestration within the vital organs. In some case were sequestration occurs in the brain this will lead to the cerebral form of malaria. Each *Plasmodium falciparum* has multiple versions of PfEMP1 with which it can alter its appearance by changing to another PfEMP1 when the immune system begins to create antibodies for the original PfEMP1 in a process known as antigenic variation. Changing of adherence molecules also means a change in the receptor on the epithelial. The change in receptor is hypothesized to possibly change the disease outcome.

RIFIN, repetitive interspersed family, is considered the most abundant multigene family. PfEMP1 along with RIFIN is considered a crucial cornerstone for the virulence of *Plasmodium falciparum* mainly due to its ability to avoid immune response through antigenic variability. RIFIN is also presented on the outer membrane of a parasite infected erythrocye as an adherence factor.

Rosettes are uninfected red blood cells that form clumps with Malaria-infected erythrocytes. Clumping occurs when particularly sticky PfEMP1 attach to other red blood cells. Only a minority of *P. falciparum* actually creates rosettes, but when they do they are known to be linked to severe malaria.

Malaria pigment (hemozoin) is released during erythrocyte rupture, causing the uncomplicated symptoms of malaria such as chills and fever.

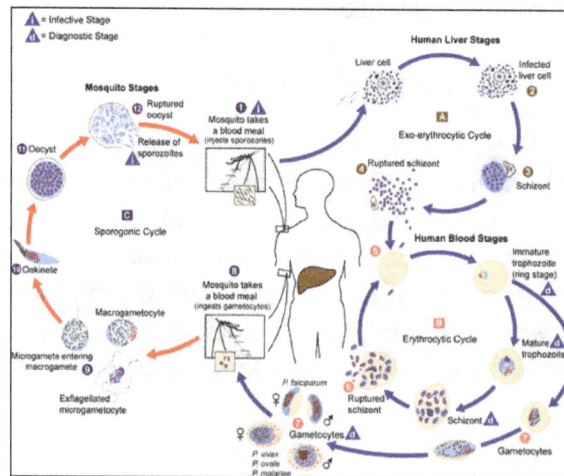

CDC illustration of the life cycles of malaria parasites, Plasmodium spp.

Life Cycle

The life cycle of malaria is a complex process infecting two hosts, human and mosquito. The process begins when an infected mosquito transferring saliva as well as sporozoites into an individual's circulatory system. These sporozoites travel to the liver and invade hepatocytes. In the liver asexual reproduction occurs through exoerythrocytic schizogony, which produces merozoites that are released back into the blood. From here the merozoites invade erythrocytes and begin the trophic period. During this period the trophozite enlarges followed by multiple rounds of asexual nuclear division to a schizont. Merozoites bud from the schizont and eventually rupture the erythrocyte releasing toxins that cause the simple symptoms of malaria, fever and chills. Merozoites eventually invade another erythrocyte, which begins another round of the blood stage replication. Some erythrocytes change into gametocytes capable of doing sexual reproduction. These cells do not lyse but instead are taken up but the next mosquito that bites, infecting the mosquito and possible more people.

In severe cases caused by P. falciparum the infected erythrocytes do not rupture, but instead use their virulent nobs to sequester into the tissue. Due to this symptoms can be vastly more complicated leading possibly lethal symptoms.

Sickle Cell Resistance

Sickle cell individuals have shown to rarely contract malaria. Research has shown that this is partially due to weakened binding of parasite-infested sickle cell erythrocytes to microvasculare endothelial cells when compared to normal hemoglobin parasite erythrocytes binding. The virulence factor PfEMP1 that normally conducts cytoadherence is altered creating a weekend attachment between it and the epithelial wall. Due to the ability to attach lacking, sequestration would also not occur limiting the severe malarial response. The mechanism for how this is done is still unknown and needs further research.

Clinical Features

Symptoms

P. falciparum normally proceeds in two forms, uncomplicated or severe. In most occurrence the severe case is observed showing symptoms such as cerebral malaria, which cause abnormal behavior, seizures, coma, or impairment of consciousness. Severe symptoms also present anemia due to destruction of red blood cells, hemoglobinuria, acute respiratory distress, low blood pressure, acute kidney failure, metabolic acidosis, and hypoglycemia. These are all due to organ failure and abnormalities in patient's blood or metabolism.

During a rare uncomplicated infection, symptoms appear flu-like. The attack lasts roughly 6-10 hours presenting a cold stage, hot stage, and sweat stage. During these stages one shows symptoms of fever, chills, sweats, headache, nausea, vomiting, body ached, and malaise.

Morbidity and Mortality

In 2010 malaria was diagnosed for 219 million people and killed 660,000 people. Roughly 70% being 5 years or younger and 75% of these cases were caused by *P. falciparum*.

Pregnant women are at higher risk for a more severe reaction to themselves and the fetus.

Young children are also at higher risk for more severe infections due to the immaturity of their immune systems.

Diagnosis

Rapid and accurate diagnosis using microscopic examination of blood smears is the most precise way to determine *Plasmodium falciparum* as the disease. CDC provides various references for microscope diagnosis along with serology, PCR, and drug resistance testing. Each species of P. falciparum has distinctive characteristics that can be seeing under a microscope. In only early form, trophozites and gametocytes of P. falciparumare seen in the blood as ring form inside the erythrocyte. There are normally multiple parasites in one erythrocytes appearing as several dots.

Treatment

The best line of defense against any form of malaria is preventative treatment, antimalarial, taken properly before, during, and after exposure to parasite.

Treatment of *P. falciparum* depends on severity of infection as well as location where the infection took place. Treatment can also vary due to an individual's age, weight, and pregnancy status.

In uncomplicated malaria, the first line of defense includes Artemisinin-based combination therapy (ACT). ACTs are used to improve treatment by overcoming the resistance by using more than one derivative of Artemisinin. The choice of which ACT to use depends on the region in which the infection took place. This is due to the varying level of resistance found in different areas. Non-ACTs such as sulfadoxine-pyrimethamin with chloroquine can also be used but are considered to have a limited sufficiency due to drug resistance.

In severe malaria, the main focus is to keep the patient from dying. Rapid clinical assessment and confirmation is key.

Prevention

Risk Avoidance

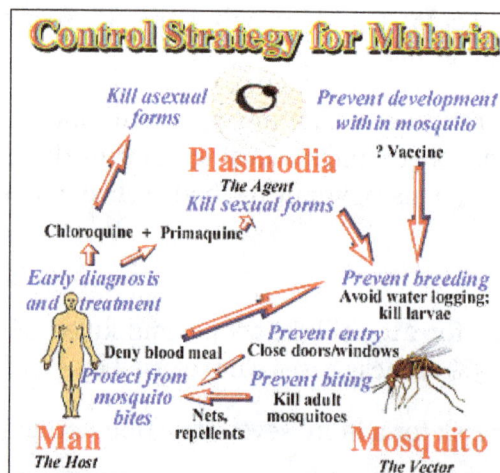

Control Strategy for Malaria

In 1992, the W.H.O. redirected the strategy towards malaria from vector control to treatment. The control of malaria entails 3 living beings: human, mosquito, and vector. Each has its own complications and if treated properly ability to stop the cycle of malaria. For successful malaria control it is now believe to target man first, mosquitoes next followed by the parasite. The web of interact allows the control of malaria on one of these systems to complement the others.

Immunization

Studies are still on going for an immunization for *Plasmodium falciparum*. The prospects of a vaccine do not look promising due to infected individuals never developing sterilizing immunity. The parasite has an impressive ability to avoid and suppress the immune system never allowing it to create the proper antibodies to fight the infection.

Host Immune Response

A key feature to the virulence of *Plasmodium falciparum* is antigenic diversity. This is the parasites ability to switch erythrocyte- associated antigens, thus evading the immune system. The alteration between antibodies only occurs in the trophozoite/schizont stage in the erythrocyte. The switches in erythrocites surface antigens as well as multiple strains of P. falciparum have prevented the creation of a vaccine. Plasmodium falciparum's clever avoidance of the spleen and immune system has created a parasite almost impossible to eradicate from the host without anti-malaria medications.

Cryptosporidium

Cryptosporidium is a protozoan parasite of medical and veterinary importance that causes gastroenteritis in a variety of vertebrate hosts. Several studies have reported different degrees of pathogenicity and virulence among Cryptosporidium species and isolates of the same species as well as evidence of variation in host susceptibility to infection. The identification and validation of Cryptosporidium virulence factors have been hindered by the renowned difficulties pertaining to the in vitro culture and genetic manipulation of this parasite. Nevertheless, substantial progress has been made in identifying putative virulence factors for Cryptosporidium. This progress has been accelerated since the publication of the Cryptosporidium parvum and C. hominis genomes, with the characterization of over 25 putative virulence factors identified by using a variety of immunological and molecular techniques and which are proposed to be involved in aspects of host-pathogen interactions from adhesion and locomotion to invasion and proliferation. Progress has also been made in the contribution of host factors that are associated with variations in both the severity and risk of infection.

Our understanding of infectious diseases has been deeply influenced by the fathers of microbiology, Robert Koch and Louis Pasteur, who independently demonstrated the relationship between the microbial world and disease. Their work and rationale led directly to the identification and characterization of the etiological agents of the world's major infectious diseases. Subsequent investigators, however, noted that the causal relationship between microbes and disease can be less straightforward than originally envisaged by Koch and Pasteur, with a variety of host and pathogen factors affecting the occurrence and outcome of disease, particularly in cases of opportunistic infections associated with a compromised host immune system. In considering this dichotomy, Isenberg reflected in 1988 that while "doubt about the meaning of pathogenicity and virulence seems inappropriate, if not ridiculous," the traditional vocabulary associated with infectivity and virulence was nevertheless insufficiently well-defined to accommodate contemporary observations of clinically important infectious diseases. Despite the work of Isenberg and others to clarify and unify a set of appropriate descriptive terminology in this area, some confusion persists, undermining contributions from the major technological advances of our time: genome sequencing, real-time PCR, DNA microarrays, transcriptomics, genetic manipulation, RNA silencing, monoclonal antibodies, fluorescence imaging techniques, and proteomics. This confusion correlates with the fact that the majority of the virulence factors from the microorganisms responsible for the world's most prevalent diseases remain poorly defined and uncharacterized.

Contemporary Description of Virulence

Virulence is commonly defined simply as the ability of a microorganism to cause disease. Virulence and pathogenicity are often used interchangeably, but virulence may also be used to indicate the degree of pathogenicity, where pathogenicity is used solely to describe the ability of a pathogen to inflict damage to the host. Virulence is commonly used to describe the likelihood of an infected person becoming ill as well as the severity of symptoms. Thus, virulence could be considered an increased risk of infection and/or an increased severity of illness, and in light of this, Woolhouse and colleagues defined virulence as, "the direct or indirect reduction in host fitness attributable to pathogen infection." Pathogen-centered views of virulence assert that pathogens are distinguished from nonpathogens by their expression of virulence factors. Although this concept appears to apply to some microbes that cause disease in healthy hosts, it has been pointed out that this does not apply to microbes that, like Cryptosporidium, cause serious disease primarily in immunocompromised hosts. While earlier definitions of pathogen-centered pathogenicity and virulence are understandable, the modern consensus is that neither microbe nor host characteristics can independently cause disease. Virulence, despite being a microbial characteristic, can exist only in a susceptible host and depends on the context and nature of the host-microbe interaction. Thus, a qualified description of virulence with which we concur can be phrased as a pathogen property which can be modulated by host susceptibility and resistance and which causes damage to the host. This description takes into account the complexity of host-pathogen interactions and the contribution of both parties to virulence and uses host damage to define virulence.

The view that virulence is a single characteristic, however, is difficult to reconcile with the fact that host-pathogen interactions are continuous and subject to changes in host, microbial, and exogenous factors. In fact, virulence is usually multifactorial, involving a complex interplay between the parasite and the host. Various host factors, including age, sex, and the status of the immune system, affect the outcome of the host-parasite interaction. In addition, the genotypic and phenotypic characteristics of the parasite define intrinsic diversity in isolate pathogenicity and virulence. Therefore, an integrated view of microbial pathogenesis and virulence accounting for the contribution of both host and pathogen factors and accommodating a spectrum of host-pathogen interactions provides a more accurate perspective.

Each of the microbiological attributes that contribute to virulence can, in general, be linked to specific structural elements or biochemical compounds within the organism; these are generally termed virulence factors. Although the terms "virulence determinants" and "virulence factors" are widely used to describe traits contributing to pathogenicity, a subtle distinction exists between the two terms. Virulence factors are "microbial traits that promote host damage", and more precisely, a virulence factor is a gene product necessary but not sufficient to cause disease. In this context, virulence factors can be defined as "contributory virulence factors", which are not singular determinants of virulence but are able to influence the severity and duration of the disease; therefore, the mutation of such virulence factors should still allow the pathogen to cause disease. This can be illustrated by the significantly reduced (but still present) virulences of a variety of mutant strains. Conversely, virulence determinants are "the factors present in a microorganism that are responsible for the relative capacity of a parasite to cause damage in a host", although perhaps a more useful, specific definition of a virulence determinant could be a gene enabling an organism to colonize the host successfully and which may then result in host pathology. An operational definition used for the identification of virulence determinants in experiments is a gene belonging to a pathogen whose inactivation or deletion leads to a loss of virulence of the pathogen and whose genetic reintroduction restores virulence. Consistent with this, virulence determinants are most straightforwardly defined as "requisite virulence factors". The identification of virulence factors and virulence determinants for Cryptosporidium spp. is not straightforward and is even more complicated in particular settings such as host immune disorders.

On a practical note, Edberg stated, "For a microbe to generate disease, a number of sequential virulence factors must be active. While clearly the genes that code for virulence must be present in the microbe, disease generation is a phenotypic phenomenon." An implication of this is that the presence of a virulence gene does not necessarily mean that it will be expressed or active, so the possession of a virulence factor gene does not necessarily equate with virulence per se. It is important to take this into account when considering appropriate methodologies for the detection and tracking of virulent strains. For instance, in many cases, genotype testing by PCR cannot definitively determine whether a strain is virulent but merely that it encodes a virulence determinant and thus is potentially virulent. In such cases, however, expression-based assays such as immunofluorescence assays (IFAs), enzyme-linked immunosorbent assays (ELISAs), or, indeed, mass spectrometry may be more appropriate alternatives, better able to discriminate a virulent phenotype.

Virulence factors are likely to be involved in adhesion, colonization, invasion, and host immune evasion. When characterized, most factors share one or more of the following properties:

- They are externally exposed, either on the surface of the parasite or as secreted proteins;

- They are hypervariable between isolates;

- They are encoded telomerically or subtelomerically;

- They are multicopy or belong to gene families; and

- They are glycosylated and/or lipoylated.

Virulence factors have long been considered important microbial traits, and decades of research have been dedicated to the identification of such factors. Such efforts have been rewarded by the discovery of many and diverse microbial genes/molecules that mediate damage and disease in infected hosts. In bacteria, virulence factors are frequently located in pathogenicity islands and can be transferred via plasmids or lysogenic bacteriophages. Classical bacterial virulence factors include toxins, fimbriae, and flagella, while in single-celled eukaryotic pathogens; perhaps the best-characterized virulence factors are hypervariable surface proteins such as those that enable antigenic variation. Other surface virulence factors of protozoa also have well-established roles in adhesion, cell invasion, resistance to the host immune response, intracellular survival, and nutrient uptake, while secreted protozoan virulence factors can vary considerably in character, from cytotoxic proteases, to signaling molecules capable of dysregulating homeostasis, to molecules involved in quorum sensing. In the postgenomic era, the discovery and validation of protozoan virulence factors in particular have been accelerated by the application of technological advances, including comparative genomics, transcriptomics, microarrays, and reverse genetics including gene replacement and small interfering RNA (siRNA).

When considering how each individual virulence factor contributes to an overall virulence phenotype, it is crucial to identify reliable measures of virulence. The ability of a microbe to cause disease in an animal model, which is central to Koch's postulate, has been the cornerstone of the measurement of virulence, but this relies on the availability of a susceptible experimental animal model. Some commonly used measures of virulence are mortality, morbidity, microbial burden, weight loss, condition, and the lifetime reproductive success of infected hosts versus uninfected hosts. Increasingly, additional measurements of virulence, such as different measures of host cell damage (both in vitro and in vivo) and the magnitude and type of inflammatory and immune responses elicited by the pathogen, are being introduced. The advent of such parameters provides researchers with useful laboratory-based and epidemiological assessments with which to consider the contribution of virulence factors to strain pathogenicity in cases where access to an animal model is not appropriate or feasible.

Cryptosporidium Virulence Factors

Several studies have tried to determine the factors responsible for the initiation, establishment, and perpetuation of Cryptosporidium infection. Cryptosporidium does not normally cause a systemic infection or penetrate deep tissue; rather, the parasite establishes itself in a membrane-bound compartment on the apical surface of the intestinal epithelium. Nevertheless, it causes significant abnormalities in the absorptive and secretory functions of the gut. This damage could be the result of direct injury to the host epithelial cells or could be indirect through the effect of inflammatory cells and cytokines recruited to the site of infection.

Virulence factors are considered to be the processes and substances by which the parasite initiates and maintains disease in the host; these factors can affect the host at any time during the life cycle from the time when the parasite enters the body until it is killed or completes the cycle and exits the host. To date, Cryptosporidium-specific virulence factors have not been characterized to the point of unequivocally establishing their roles in causing damage to the host or proving that their deletion or inactivation results in a decrease of disease severity. This is mainly because, unlike other apicomplexan parasites such as Toxoplasma and Plasmodium, it remains difficult to employ in vitro cultivation and reverse genetics techniques with this parasite, meaning that genes cannot be readily knocked out or knocked down in experiments designed to examine virulence by a straightforward loss or gain of pathogenicity.

Putative virulence factors for Cryptosporidium have been identified as genes involved in the initial interaction processes of Cryptosporidium oocysts and sporozoites with host epithelial cells, including excystation, gliding motility, attachment, invasion, parasitophorous vacuole formation, intracellular maintenance, and host cell damage.

Adherence Factors

A critical initial step in establishing infection is parasite attachment to host cells. Two classes of proteins, namely, mucin-like glycoproteins and thrombospondin-related adhesive proteins, have been characterized and shown to mediate adhesion.

CSL (circumsporozoite-like glycoprotein), with a molecular mass of ~1,300 kDa, was described by Riggs and colleagues and is associated with the apical complex of sporozoites and merozoites. CSL is released as a soluble glycoprotein and contains a ligand that binds specifically to a receptor on the surface of human and bovine intestinal epithelial cells. The zoite ligand was shown to be involved in attachment and invasion. Monoclonal antibodies to CSL elicited changes in sporozoites and merozoites, similarly to the malarial circumsporozoite precipitate (CSP) reaction, and caused the complete neutralization of sporozoite infectivity.

Glycoprotein 900 (gp900) is a large glycoprotein identified by the immunoprecipitation of sporozoite extracts with hyperimmune bovine colostrum. This large mucin-like glycoprotein is located in micronemes and at the surface of invasive merozoites and sporozoites. gp900 is deposited in trails during gliding motility and is known to mediate invasion. The deduced amino acid sequence of gp900 has a signal peptide and a transmembrane domain. Specific antibodies to gp900 can competitively inhibit infection *in vitro*. gp60 N- and C-terminal peptides include a hypothetical signal sequence and a glycosylphosphatidylinositol (GPI) anchor attachment site, which are highly conserved among all *Cryptosporidium* isolates, suggesting that both features are important. The rest of the gene has a high degree of polymorphism, which is far greater than any other *Cryptosporidium* protein-encoding locus. A typing technique based on the sequence analysis of this polymorphic sporozoite surface glycoprotein has been extensively used and allowed the classification of *Cryptosporidium* isolates into gp60 subtypes. Interestingly, some gp60 subtypes were associated with strain virulence. Cama and colleagues reported differences in clinical manifestations between *Cryptosporidium* species and subtypes in HIV-infected persons and in children. Similarly, a study in China showed that two subtypes were associated with extended outbreaks in hospitalized children.

A sporozoite and merozoite cell surface protein gp15/40/60 complex has been described (Cpgp40/15 in *C. parvum* and Chgp40/15 in *C. hominis*): Strong and colleagues reported that gp15/40/60 mRNA is translated into a ~60-kDa glycoprotein precursor during the intracellular stages of the *C. parvum* life cycle. Independently, Cevallos and colleagues cloned and sequenced the same gene from *C. parvum* genomic DNA and called it Cpgp40/15. Shortly after synthesis, the 60-kDa precursor is proteolytically processed to generate 15- and 45-kDa glycoproteins. gp40 is localized at the surface and the apical region of the parasite and is shed from the surface, while gp15 is on the surface of sporozoites and is shed in trails during gliding movement. Proteins present in the trails of gliding zoites have been shown to play a role in parasite attachment, invasion, and motility. In *C. parvum*, gp15 is attached to the membrane via a GPI anchor. Both gp40 and gp15 display O-linked α-*N*-acetylgalactosamine (α-GalNAc), which is thought to be involved in invasion and attachment, since lectins that recognize these determinants block sporozoite attachment. A monoclonal antibody against gp15 neutralized infectivity *in vitro* and passively protected against the disease *in vivo*. It was suggested that a protein complex of gp900/gp40/p30 is formed to mediate adhesion via lectin activity.

P23 is a 23-kDa sporozoite surface protein that is antigenically conserved across geographically diverse isolates and is deposited in trails during the initial stages of infection. P23 elicits antibody responses in animals and humans exposed to *C. parvum*. P23 has neutralization-sensitive epitopes, and monoclonal antibodies were found to significantly reduce infection in mice and protect calves against cryptosporidiosis.

Of the micronemal proteins (MICs), TRAP-C1 (thrombospondin-related adhesive protein *Cryptosporidium*1) is a 76-kDa protein localized on the apical pole of sporozoites. It showed sequence and structural homology to members of the thrombospondin family of adhesive proteins in other apicomplexan parasites (*Plasmodium* spp., *Toxoplasma gondii*, *Eimeria tenella*, and *Neospora* spp.). TRAP and structurally related proteins are involved in parasite gliding motility and cell penetration. Putignani and colleagues characterized "CpTSP8," from a family of 12 *C. parvum* thrombospondin-related proteins (CpTSP2 to CpTSP12), which were identified by using bioinformatic tools. Those authors showed that CpTSP8 is located at the apical complex of sporozoites and merozoites and is translocated onto the parasite surface, as is typical of MICs. Therefore, CpTSP8 was renamed CpMIC1. MIC proteins have been shown to be essential for host cell attachment/invasion and gliding motility.

gp900, gp40, gp15, Cpa135, Cp2, P23, and TRAP-C1 have or are predicted to have mucin-type O-glycosylation, suggesting that this type of posttranslational modification is common in proteins involved in attachment and invasion. Because mucin-like proteins were shown to be important for host-parasite interactions for *Cryptosporidium*, O'Connor and colleagues undertook data mining of the *Cryptosporidium* genome databases to identify other mucin-like genes. They discovered a single locus of seven small mucin sequences (CpMuc1 to CpMuc7), which were expressed throughout the intracellular development stages. Specific antibodies inhibited infection *in vitro*, which is consistent with a role in host cell invasion.

The proteolytic processing of surface and apical complex proteins by parasite proteases has been shown to be required for the invasion of host cells and for egress from them. Further effort was focused on the identification of *Cryptosporidium* proteases. Wanyiri and colleagues characterized a *C. parvum* subtilisin-like serine protease (CpSUB) and showed that this protein is likely to be responsible for the processing of gp40/15.

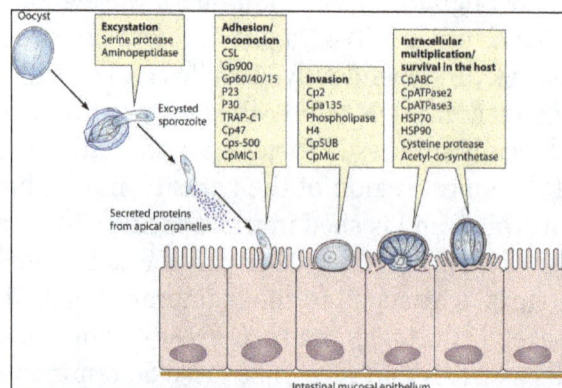

Cryptosporidium virulence factors described to date and their contribution to the parasite life cycle.

The identification of these molecules that mediate attachment to and invasion of epithelial intestinal cells is advancing our understanding of the cell biology of *Cryptosporidium* infection. These proteins have features in common with other apicomplexan proteins implicated in mediating host cell interactions and are expressed on the surface of the invasive *C. parvum* sporozoite and merozoite stages and shed in trails by gliding zoites. The relative contribution of each individual molecule remains to be determined. It is likely that by using a large number of seemingly redundant adhesive molecules, the parasite can maximize the opportunity for cell attachment across a broad range of potential hosts. It is also possible that quantitative or qualitative differences in these glycoproteins may confer selectivity for host attachment.

Cellular Damage

Cell damage in enterocyte monolayers has been documented through the disruption of tight cell junctions, a loss of barrier function, the release of lactate dehydrogenase, and increased rates of cell death. The mechanisms that cause cellular damage during *Cryptosporidium* infection remain unknown; however, several molecules can cause direct tissue damage, such as phospholipases, proteases, and hemolysins.

Proteases have important functions in a parasite's life cycle, such as mediating protein degradation, the invasion of host tissues, and the evasion of host immunity. Distinct protease activities have been identified for *Cryptosporidium* sporozoites: aminopeptidase, cysteine protease, and serine protease activities have been implicated in the excystation process. The identification of functional proteases in sporozoites during excystation and the prevention of infection in the presence of protease inhibitors suggest that proteases are important in the initial stages of *Cryptosporidium* infection.

Hemolysin H_4 has been identified by the screening of a *C. parvum* expression library on sheep blood agar. H_4 has sequence similarity to the hemolysin of enterohemorrhagic *Escherichia coli* O157:H_7. The function of H_4 is unknown, but its ability to disrupt cell membranes suggests a role in cellular invasion and/or the disruption of vacuolar membranes, which would allow merozoites to exit the parasitophorous vacuole and spread to adjacent cells. Another *Cryptosporidium* protein of interest is a *C. parvum* ATP-binding cassette (ABC) transporter gene (CpABC) localized in proximity to the electron-dense feeding organelle of the parasitophorous vacuole, which may well be associated with meeting key nutritional requirements. Interestingly, these genetic elements share structural similarities with bacterial genes, which are critical in producing secretory diarrhea.

Heat Shock Proteins

Heat shock proteins (HSPs) are a family of large conserved proteins. They are usually defined by their apparent molecular weights determined by sodium dodecyl sulfate-polyacrylamide gel electrophoresis (SDS-PAGE), with HSP90, HSP70, and HSP65 being common families. The level of synthesis of HSPs, especially HSP70, increases dramatically under stressful conditions (sudden shifts in temperature, decreased availability of nutrients, and immune attack). HSPs function as intracellular chaperones for other proteins; play an important role in protein-protein interactions; and facilitate the transport, folding, assembly, biosynthesis, and secretion of newly formed proteins.

Considerable polymorphism in the HSP70 gene has been identified and was used for genotyping purposes. However, HSPs are under selective pressure, and their high degree of polymorphism might not reflect the genetic relationships among isolates or subtypes. For the closely related apicomplexan parasite *T. gondii*, it was demonstrated that quantitative and qualitative differences in HSP expression levels are related directly to parasite virulence. High levels of expression of HSP70 were detected in virulent strains of *T. gondii* grown in mice, but little expression of HSP70 was observed in avirulent strains. The relationship between the level of HSP expression and *Cryptosporidium* virulence warrants further investigation.

Contingency Genes

Contingency genes, as opposed to conserved housekeeping genes, are highly variable. They are common in pathogenic microbes, including viral, bacterial, fungal, and protozoan pathogens. Contingency genes are subject to spontaneous recombination rates higher than the background rate that applies to the other genes in the genome. These mutational events allow rapid switches in phenotype that are conducive to survival and proliferation in the host. In eukaryotic pathogens, these contingency genes are often associated with telomeres. Telomeres are prime sites for these quickly evolving genes that mediate host-parasite interactions. Examples of contingency virulence genes include the variant surface glycoprotein (VSG) in *Trypanosoma brucei*, the *var* genes in *Plasmodium falciparum*, the *trans*-sialidases of *Trypanosoma cruzi*, the major surface glycoproteins (MSG) of *Pneumocystis carinii* and *Pneumocystis jirovecii*, and the subtelomeric variable secreted proteins (SVSPs) of *Theileria* sp.

Comparative genomic analyses of *C. parvum* and *C. hominis* isolates showed that the most highly divergent regions are located near the chromosome ends. The majority of such genes either were transporters or contained predicted signal peptides, a result which supports a role for telomeric contingency genes in dictating host specificity. Our comparative genomic analyses of *C. hominis* and *C. parvum* allowed the identification of a new family of telomerically encoded *Cryptosporidium* proteins: the *C. parvum*-specific protein (Cops-1) and the *C. hominis*-specific protein (Chos-1). *Cops-1* and *Chos-1* appear to be distantly related and share similar characteristics, including having a telomeric location, encoding secreted glycoproteins of 50 kDa, and containing clear but distinct internal repeats. The proteins that they encode are predicted to be highly glycosylated, as shown in figure. These characteristics suggested a possible role in host-parasite interactions, and we undertook a further characterization of the Cops-1 protein. Consistent with such a role, Cops-1 appears to be a secreted protein localized to the oocyst content and sporozoite surface of *C. parvum* but not *C. hominis*. Additionally, sera from *C. parvum*-infected patients recognized a 50-kDa protein in antigen preparations of *C. parvum* but not *C. hominis*, consistent with Cops-1 showing species-specific expression and being antigenic for patients.

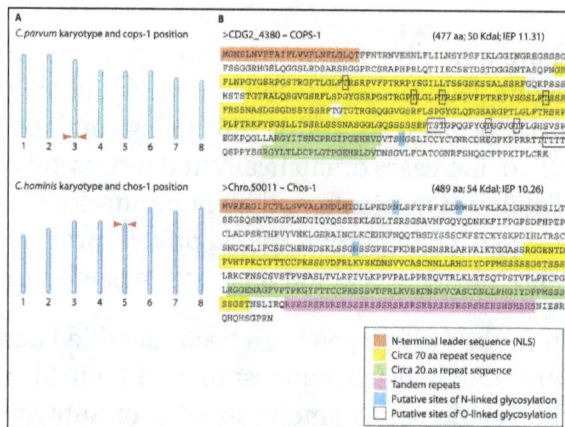

Characteristics of the new family of telomeric *Cryptosporidium* proteins, Cops-1 and Chos-1.

In figure, (A) Genomic positions of the Cops-1 and Chos-1 genes. (B) Predicted Cops-1 and Chos-1 proteins are 50 kDa, secreted, and serine rich; contain internal repeats; and are highly glycosylated. aa, amino acids; IEP, isoelectric point.

Host Factors and Cryptosporidium Virulence

Host factors are critical determinants of the outcomes of host-pathogen interactions. It has been demonstrated that the severity of *Cryptosporidium* infection is highly dependent on host factors. The host immune capacity is the most important host factor affecting both the probability of an infection and the severity of the subsequent disease. This observation is most obvious when considering the impact of immune suppression on disease courses. Petry and colleagues presented a comprehensive review of both innate and adaptive host immune responses to *C. parvum* infection. As the relevance and individual contribution of each of these responses remain unclear, the effect of any particular immunosuppressive disorder on the severity of *Cryptosporidium* infection remains difficult to predict. What is clear, however, is the link between an increased severity of disease and certain types of immune suppression. Indeed, many of the earliest-identified cases of human cryptosporidiosis were immunosuppressed individuals.

The severity of illness depends on the degree of the immunocompromised status. In immune-suppressed subjects, cryptosporidiosis is no longer self-limiting and can be life-threatening. However, not all forms of immune suppression lead to an increased disease severity of cryptosporidiosis. The main risk seems to be immune-suppressive disorders that impact T cell function, the most obvious being HIV/AIDS. In AIDS patients, the most severe disease occurs in people with a CD4 cell count of less than 50. Flanigan and colleagues reported that in AIDS patients, self-limited cryptosporidiosis was seen in patients with higher CD4 counts. Further support for the role of T cell function was shown by improvements in the symptoms of cryptosporidiosis following treatment with highly active antiretroviral therapy (HAART).

Of the primary immune deficiencies, those diseases that have been most clearly linked to an increased severity of cryptosporidiosis have included severe combined immunodeficiency syndrome, X-linked hyperimmunoglobulin M syndrome, and CD4 lymphopenia. All three of these deficiencies are also characterized by impaired T cell function. In contrast, immune deficiencies affecting B cell function have not been so associated, although one report linked chronic intractable diarrhea and significant weight loss associated with cryptosporidiosis in an infant with a gamma interferon deficiency. There is also evidence that patients with malignant disease may also be more susceptible to infection, especially following bone marrow transplantation. However, the number of reports of *Cryptosporidium* in patients with malignancy is rather lower than for people living with HIV, probably reflecting a less common exposure in patients with malignancy than in patients with HIV infection. There have also been reports of cryptosporidiosis in patients following solid-organ transplantation, although data were insufficient to make an adequate risk assessment.

Acquired immunity would also appear to be important in preventing illness, as illustrated by reports of waterborne outbreaks of disease. During one such outbreak in Talent, OR, symptomatic cases of cryptosporidiosis occurred only in visitors to the town and not in residents, suggesting that residents had acquired immunity from previous exposures. For another outbreak, it was shown that swimming in a pool (a known risk factor) was negatively associated with disease in a drinking water-associated outbreak. One study showed that people with high anticryptosporidial antibody levels (IgG) were less likely to report diarrhea. What remains unclear is the duration of such acquired immunity. Evidence from adult human feeding experiments suggests that immunity may not last much longer than a year or so, and the protection acquired translates into less severe clinical symptoms. However, this is unlikely to be an all-or-nothing effect, as it appears that immunity merely shifts the dose-response curve to the right. In other words, even people who have acquired immunity can become ill if the ingested dose is sufficient; indeed, infection and diarrhea were reported in these subjects at a dose 20-fold higher than that for seronegative volunteers. There is also some evidence that such partial immunity may give rise to symptoms even though the oocysts remained undetected in stool samples. That same study showed that oocyst shedding occurred in only 53.8% of subjects with clinical cryptosporidiosis.

An interesting aspect of the issues around immunity is whether regular exposure to low doses of *Cryptosporidium* may actually have benefit. This notion was first raised independently by Craun and colleagues and Hunter and Quigley, who reported that attack rates were significantly higher for outbreaks associated with groundwater than surface water consumption. They argued that people who use surface water sources were regularly exposed to small numbers of oocysts and thus did not experience many outbreaks, unless there was a major breakdown in treatment. The role of immunity in modulating virulence is also supported by a risk factor analysis which identified the eating of raw vegetable as a protective factor. Using mathematical modeling, Swift and Hunter showed how an increased probability of exposure could lead to earlier infection but reduced risk later in life, a hypothesis supported by data from seroepidemiology studies. Indeed, if, as recently hypothesized, repeated exposure serves to maintain an otherwise short-lived immunity, there would be an optimal daily probability of exposure that would minimize illness rates; however, any increased exposure would increase the risk to the youngest individuals, who are likely to be the most vulnerable. In a study modeling the risk of cryptosporidiosis associated with unreliable drinking water supplies in Africa, Hunter and colleagues argued that the main impact of increased exposure would be an earlier age of first infection.

The age of the host is a major factor in the epidemiology of infection, with children being most at risk for cryptosporidiosis, as shown by the cumulative data for cryptosporidiosis cases detected between 1989 and 2008. Illness in children is not necessarily more or less severe than that in older age groups, although they are more likely to report vomiting. In a case-control study, 70% of children (<17 years old) reported vomiting, compared to only 46% of adults (>16 years old) (P. R. Hunter, unpublished data). Infection maybe subclinical in the young and asymptomatic carriage in children has been suggested to be an important reservoir for *C. hominis* infection. The high incidence of cryptosporidiosis in children probably reflects a lack of immunity due to few prior exposures and the immaturity of the gut mucosa.

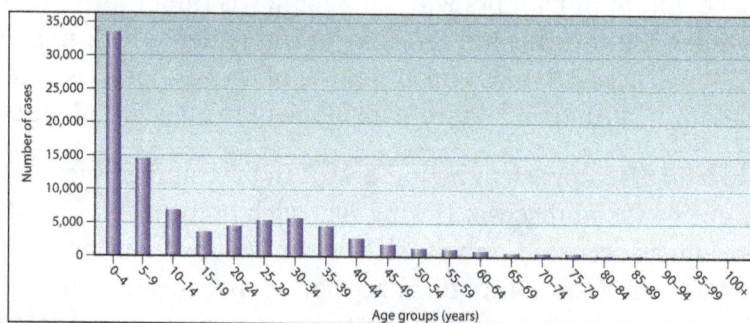

Total number of cryptosporidiosis cases detected in England and Wales by age group between 1989 and 2008.

Of particular relevance in developing countries is the issue of malnutrition in children. Cryptosporidiosis is common in children, and it is frequently associated with persistent diarrhea, malnutrition, stunted growth, and cognitive impairment. There is good evidence from animal studies that cryptosporidiosis exacerbates malnutrition and that cryptosporidiosis is more severe in malnourished mice. For humans, the evidence also supports this synergistic relationship. However, the exact mechanisms behind the increased severity of illness in malnourished children are not yet fully clear. One likely explanation is that in a malnourished child, the damage to the gut mucosa with a new *Cryptosporidium* infection would lead to further impaired nutrient absorption and stunting that would not be such an issue for a well-nourished child.

Host behavioral factors also play an important role in the variation in the risk of cryptosporidiosis by increasing or decreasing the risk of exposure to the parasite. Several such factors are discussed above and include foreign travel, contact with livestock, swimming in a swimming pool, caring for a case of infection, changing diapers, having multiple sexual partners, having pets, and using public toilets. These factors are likely to affect susceptibility to infection, but they are unlikely to affect infection courses or modify clinical manifestations.

Chapter 3

Plant Parasites

A plant which derives some or all of its nutritional requirement from another plant is known as a parasitic plant. Some of the common parasitic plants are Dodder, Rhinanthus minor, Viscum album and Rafflesia Arnoldii. All the diverse aspects of parasitic plants have been carefully analyzed in this chapter.

Parasitic Plant

Parasitic plant is a plant that obtains all or part of its nutrition from another plant (the host) without contributing to the benefit of the host and, in some cases, causing extreme damage to the host. The defining structural feature of a parasitic plant is the haustorium, a specialized organ that penetrates the host and forms a vascular union between the plants.

Parasitic plants differ from plants such as climbing vines, lianas, epiphytes, and aerophytes; though the latter are supported by other plants, they are not parasitic, because they use other plants simply as a structure on which to grow rather than as a direct source of water or nutrients. Another group of plants that is sometimes confused with parasites is the mycoheterotrophs. Similar to parasitic plants, mycoheterotrophs may lack chlorophyll and photosynthetic capacity, but they live in symbiotic association with fungi that gain nutrition from autotrophic (self-feeding) plants or decaying vegetation. Such plants are not classified as parasitic, because they do not appear to harm the fungi and they lack haustoria.

All parasitic plant species are angiosperms, among which parasitism has evolved independently about 12 times. Some examples of parasitic angiosperm families include Balanophoraceae, Orobanchaceae, and Rafflesiaceae. Although one species of gymnosperm, Parasitaxus usta, have been proposed to be parasitic, it actually may be a mycoheterotroph as it appears to involve a fungal symbiont.

Branched Broomrape: Branched broomrape (Orobanche ramose), a parasitic plant. Frequently attacking agricultural crops, including tomatoes and tobacco, the plant is an obligate parasite and requires a host for its nutritional needs.

Host Dependence

Parasitic plants evolved from nonparasitic plants and thus underwent an evolutionary transition from autotrophy to heterotrophy, which may be partial or complete. Indeed, parasitic plants differ in the extent to which they depend on their hosts for nutrients. Hemiparasites have at least some ability to photosynthesize; they primarily rely on their hosts for water and mineral nutrients. Holoparasites, on the other hand, are nonphotosynthetic and depend on their hosts for all nutrition.

European mistletoe: Numerous European mistletoe plants (Viscum album) parasitizing a tree. Mistletoes are hemiparasites, meaning they have some photosynthetic ability, and can utilize a variety of host species.

Species of parasitic plants also differ with regard to whether they need the host in order to complete their life cycle. Obligate parasites have an absolute requirement for a host, whereas facultative parasites can live and reproduce in the absence of a host. All holoparasites are, by definition, obligate. Even though hemiparasites are photosynthetically competent, some species nevertheless are obligate in terms of their reliance on a host for their reproduction.

Morphology of Parasitic Plants

Given the different origins of parasitism, it is not surprising that parasitism is manifested in diverse ways. Some species parasitize the roots of their hosts, whereas others attack stems. The haustorium itself may develop from roots or stems, depending on the parasite species, and haustoria display a wide range of morphologies. One prominent distinction is in the proportion of a parasite that grows internally versus externally to the host. For most parasite species, only the haustorium is embedded inside the host, serving to feed the parasite located externally to the host. However, the haustorium of some species proliferates in such a way that all vegetative growth occurs within the host (endophytically), and the parasite emerges only to flower. Examples of that include members of the genus Rafflesia, which grow inside the tropical vine Tetrastigma, and stemsuckers (genus Pilostyles), which live within members of the pea family (Fabaceae).

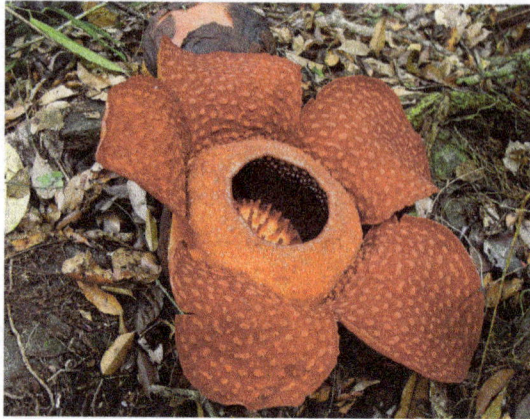

Rafflesia flower: The fetid flower of Rafflesia arnoldii, the largest known flower in the world. The plant is an endoparasite on the vines of Tetrastigma, emerging from its host only to flower.

There are many other examples of specialized morphology or life cycles that have evolved either as a necessity for parasitism or as a result of the transition from autotrophy. Holoparasite species that do not need to absorb water from the soil may have root systems that are greatly reduced in size or missing entirely. Similarly, leaves that are not needed for photosynthesis may be reduced to scales, and parasite colour may range from cream to yellow to purple, since chlorophyll production is unnecesary. An example of a holoparasite with such features is dodder (Cuscuta), which during vegetative growth has no roots and only scale leaves and therefore appears to be simply a yellow or orange stem with a network of host connections. Flowers of parasitic species often are similar to those of nonparasitic plants, although notable examples of extreme morphology include the huge Rafflesia flower and the bizarre, fleshy Hydnora inflorescence.

Hydnora flower: The unusual flower of Hydnora africana, a holoparasitic plant native to southern Africa.

Host Identification

In order to survive and reproduce, parasitic plants must be able to recognize the presence of a neighbouring plant and have mechanisms to ensure that their seeds encounter appropriate hosts. The seeds of generalist parasites (those with a wide range of potential hosts) typically germinate under environmental conditions that are favourable to nonparasites. Once the parasite seeds have developed into seedlings, they then must locate a host. The roots of generalist root parasites are able to recognize and form haustoria with the roots of other plants that they encounter in the soil. Similarly, the threadlike shoot of the generalist dodder seedling, a stem parasite, elongates and uses information about the colour of the host and volatile chemicals it produces in order to orient growth toward the host; once the shoot has reached the host plant, it coils and forms haustorial connections.

Specialist parasites, many of which are obligates, tend to have additional mechanisms to detect their specific host plants. The best-studied examples are parasites of the family Orobanchaceae (e.g., Orobanche, Phelipanche, and Striga), the seeds of which are extremely small and may sit in the soil for years until the root of an appropriate host has grown nearby. At that point the parasite seed detects a chemical signal (generally a strigolactone, a type of plant hormone) exuded from the host root, which triggers germination of the parasite seed. The parasite radicle (embryonic root) then grows a short distance; typically less than 2 mm (0.08 inch), to contact the host root and produce a haustorium.

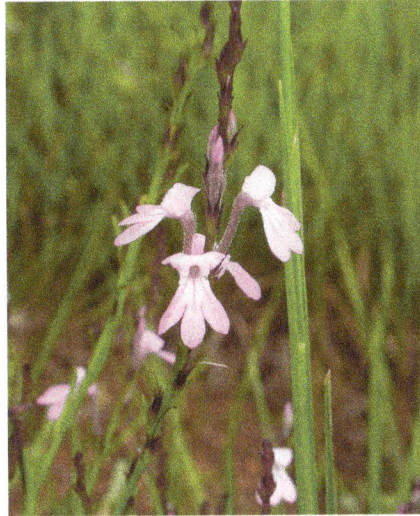

Witchweed (Striga bilabiata): An obligate root parasite. Witchweeds commonly parasitize cereal crops, and their tiny seeds can persist in agricultural fields for many years.

Impact of Parasitic Plants

Although most parasitic plant species are considered wild flowers or botanical curiosities, several species are weedy and capable of causing substantial losses of agricultural crops. Among the most harmful are members of the Orobanchaceae. Witchweeds (Striga species), for example, generally attack cereal crops, including corn (maize), sorghum, millet, and rice, in Africa. The common name of the plant reflects the dramatic effect it has on affected crops, seeming to magically arrest host growth and devastate yields. The broomrapes (Orobancheand Phelipanche species) attack broadleaf crops in the Mediterranean and Middle East and are a major constraint to the production of legumes, oilseed crops, and solanaceous crops (e.g., tomatoes, peppers, potatoes, and eggplants). Additional weedy genera include dodders, which attack a wide variety of broadleaf crops, and dwarf mistletoes (Arceuthobium), which damage coniferous trees. In all cases, the close physical connection between parasite and host, including location of the parasite belowground or within the host, makes control of parasitic weeds difficult.

Dwarf mistletoe: (Arceuthobium minutissimum) growing on a pine tree.

Parasitic Flowering Plants

Fungi, nematodes, bacteria, and viruses are probably the first things that come to mind when thinking of plant pathogens. These organisms certainly do cause damage to economically important plants, but surprisingly parasitic flowering plants are also important pathogens.

Most plants are autotrophs and produce their own carbon compounds through photosynthesis. Although some plants such as Indian pipe (Monotropa) lack chlorophyll and appear to be parasitic, they are mycoheterotrophs (parasites of mycorrhizal fungi) and, hence, only indirectly parasitize the trees on which the mycorrhizal fungi are found. Here we define a parasitic plant as an angiosperm (flowering plant) that directly attaches to another plant via a haustorium. A haustorium is a specialized structure that forms a morphological and physiological link between the parasite and host. It is useful to make a distinction between the terms "parasite" and "pathogen.

If a plant also induces disease symptoms in a host, then it is a pathogen as well as parasite. A general term that refers to both parasites and mycotrophs that derive carbon from sources other than their own photosynthesis is heterotrophic, which simply means "different feeding."

Types of Parasitic Plants

Parasitic plants can be categorized based on different criteria such as where they attach to the host, the degree of nutritional dependence upon the host, or whether they require a host to complete their life cycle. In terms of location on the host, two basic types can be distinguished: stem parasites and root parasites. Stem parasites occur in several families, and pathogenic members include some mistletoe and dodder (Cuscuta and Cassytha). Root parasites are more common and occur in diverse taxonomic groups. Some of the most economically important root pathogens are in the broomrape family, Orobanchaceae. Parasitic plants may also be classified as hemiparasites or holoparasites. Hemiparasites contain chlorophyll when mature (hence are photosynthetic) and obtain water, with its dissolved nutrients, by connecting to the host xylem via the haustorium. Holoparasites lack chlorophyll (and are thus nonphotosynthetic) and must rely totally on the contents of the host xylem and phloem. All holoparasites require a host to complete their life cycle, thus they are obligate parasites. Root hemiparasites contain chlorophyll and some (e.g. Triphysaria and Odontites) complete their life cycle without hosts. But in nature, nearly all root hemiparasites are attached to host roots. Although these definitions imply absolute and discrete categories, intermediates exist such as some Cuscuta (dodder) species that are intermediate between the hemi- and holoparasitic conditions and hemiparasitic Orobanchaceae that are intermediate between the facultative and obligate conditions.

Morphological Features

In some stem parasites, such as Cassytha and Cuscuta (dodder), the vegetative portion consists solely of a stem and scale leaves. In contrast, the casual observer may not recognize that many of the photosynthetic root hemiparasites are indeed parasites because they are green with fully formed leaves. As the degree of parasitic dependence increases (i.e. the evolution from hemiparasitism to holoparasitism), profound changes occur in the morphology of the parasitic plant. In general, holoparasites tend to have leaves reduced to scales (or absent in Hydnoraceae), succulent stems, and a primary haustorium (derived from the seedling radicle). The best examples of the evolutionary stages from hemi- to holoparasitism can be seen among various representatives of the broomrape family.

The sandalwood order (Santalales) is the most morphologically and physiologically diverse group of parasitic plants. The early diverging families in this order are autotrophic (lacking haustoria, thus nonparasitic), whereas more derived members are root parasites. In addition to mistletoes, which evolved five separate times in the order, two families (Balanophoraceae and Mystropetalaceae) are composed entirely of holoparasites.

The largest family in the sandalwood order is Loranthaceae which contains stem parasitic plants commonly known as mistletoes. The Anglo-Saxon word misteltan derives from the Old German word "mist" for dung and "tan" meaning twig. This is an apropos name given that birds defecate the seeds of mistletoes onto tree branches. The connection to birds is also reflected in the Portuguese name for mistletoe: erva de passarinho ("bird herb"). Most mistletoe are woody shrubs, often with brittle stems and leaves. However, three genera of Loranthaceae are actually root parasites and one of these (Nuytsia) of western Australia represents the first lineage to diverge in the family. Adult plants of dwarf mistletoe (Arceuthobium, Viscaceae) fix very little carbon and depend heavily upon their hosts for carbohydrates. Despite this, their seedlings are photosynthetic, thus these plants cannot be considered holoparasites.

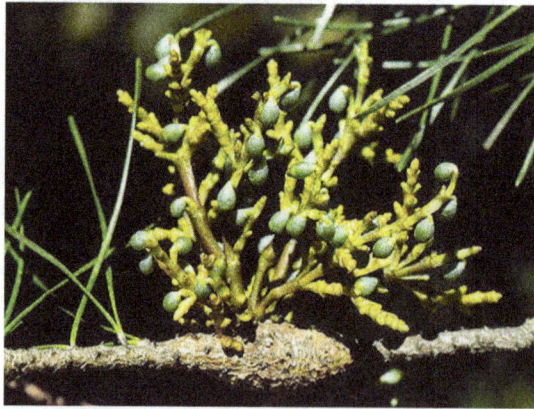

The range of flower sexual conditions (e.g. bisexual and unisexual) in parasitic plants is diverse. Dioecy (dioecious = separate male and female plants) is relatively rare in angiosperms, occurring in ca. 7% of all genera. Interestingly, dioecy is more frequent among heterotrophic plants (parasites and mycoheterotrophs) than it is among autotrophic plants. Just over 6% of autotrophic angiosperm genera are dioecious whereas over 15% of the heterotrophic genera are dioecious.

Evolutionary Development of Parasitism

What evolutionary conditions might favor the development of parasitism in plants? Root parasitism confers certain advantages, especially for annual plants. When the parasite seedling forms a haustorium, it obtains a mature, functioning root system by "assuming" the root system of its host plant. Is it energetically more cost effective for the parasite to form a haustorium on an existing root system rather than to produce one of its own? Resource allocation studies of parasites might help answer this question. Facultative hemiparasites have transpiration rates higher than their hosts, and so they are most often found in open, sunny areas. But these areas also often have dense groundcover vegetation and competition for resources in the rhizosphere is great. These conditions could have favored the evolution of root parasitism.

The stem parasitic mistletoes also exceed their hosts' transpiration rates. Given this, it is not surprising that mistletoes are most abundant in areas where access to sunlight is not limited, such as savannas and at the top of forest canopies where shading is avoided. Diversification in Loranthaceae occurred during the Oligocene, a time when temperate deciduous woodlands and grasslands were displacing tropical biomes. Within the sandalwood order, each of the five cases of the evolution of aerial parasitism (mistletoes) can be traced to root parasitic ancestors. The aerial portions of woody plants certainly represented an unexploited niche that offered opportunities for colonization by such early mistletoes.

Host Interactions

The modes of host selection and specialization of parasitic plants is extraordinarily broad. Castilleja and Cuscuta (dodder) can parasitize hundreds of different hosts in diverse families; in contrast, Epifagus virginiana (beech drops) occurs only on Fagus grandifolia (beech). The same generalization can be made about mistletoes in which some species are generalists and others specialists. Evidence exists that the generalist strategy has the greatest chance for survival over evolutionary time.

The terms host range versus host preference describe different aspects of the parasitic relationship. Host range refers to the total number of different species that can be parasitized. For example, Seymeria cassioides invariably attacks pines in nature, but in pot studies where a variety of angiosperms and gymnosperms are artificially inoculated, these plants are parasitized. Host preference, referring to the choice of the most desirable host for optimal growth, typically is much narrower (e.g., Musselman and Mann). Cuscuta (dodder) species usually have extremely broad host ranges, and can even attach to many different hosts at once. But in nature, they are found regularly on few hosts, and the parasite can often be located by first finding the preferred host.

There has been considerable research on the chemical signals between microorganisms and host roots. Accumulating evidence indicates that chemical signaling is also common in root parasites, especially germination stimulants. The seedling phase is the most vulnerable part of the life cycle because life-sustaining attachments to the host are made at this stage. In parasites with tiny ("dust") seeds with minuscule food reserves, this period is especially critical because the seedling will die in a few days without a host attachment. Parasites with larger seed reserves can survive longer. In general, parasites with tiny seeds, i.e., those less than 0.45 mm long, require a host stimulant to germinate whereas larger seeds do not.

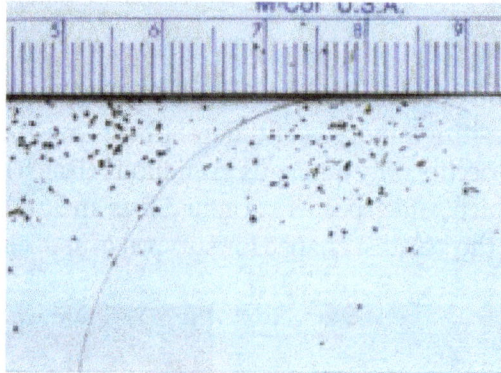

The stem parasitic dodders (Cuscuta) provide intriguing models for host selection because they have nastic movements (in response to a host stimulus) that allow them to "forage" for (or move towards) hosts. The remarkable observation has been made that Cuscuta pentagona uses volatile chemical cues to select hosts with the highest nutritional status.

Great progress is being made in understanding host resistance responses in Striga (witchweed) and Orobanche owing to advances in molecular biology. Many different avenues are being explored including genetically engineering the host crop plants to be resistant to the parasite as well as attenuating the virulence of the parasites themselves.

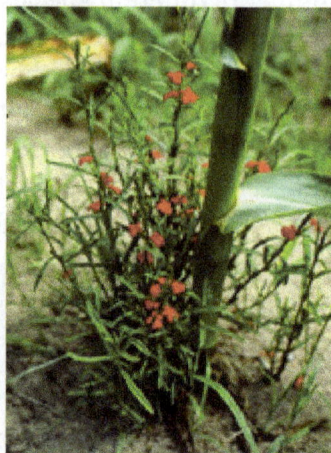

Distribution

Parasitic plants can be found in all major biomes on earth, being absent only in the coldest regions (Antarctica, North Pole). Pedicularis dasyantha (Orobanchaceae) is found on the Svalbard archipelago of the Arctic Ocean, more than 80° N latitude. Nanodea muscosa (Nanodeaceae, formerly Santalaceae) is found at the other polar extreme, Tierra del Fuego of far southern South America. Tropical rainforest biomes support high plant diversity overall, however, parasitic plants make up a relatively small proportion of the flowering plants found there. Exceptions include two groups: mistletoes and holoparasites. Being stem hemiparasites, the mistletoes (mainly Loranthaceae and Viscaceae) are found high in the canopy of forest trees where sunlight is not as limited as in the understory. On the forest floor, little sunlight penetrates, hence nonphotosynthetic holoparasites such as Balanophoraceae and Rafflesiaceae can be found here. In general, parasitic plants reach their highest diversity in biomes that are not light limited such as grasslands, savannas, xeric shrublands, and even semiarid deserts. For example, the highest number of endemic genera in Orobanchaceae are found in South Africa, followed by the Southeastern United States (open, fire-maintained communities), and the Alps of Central Europe (open grasslands). The Mediterranean climate of South Africa is also the center of diversity for the root hemiparasite Thesium (Thesiaceae, formerly Santalaceae), the most speciose genus in the sandalwood order.

Dispersal

Like many weeds, several pathogenic parasitic plants have been effectively spread by humans. Contaminated seed has spread the North American dodder (Cuscuta campestris) throughout most of the world. This species of dodder is the most widespread parasitic weed in the world, and was first described (as C. pentagona) from a major USA seaport, Norfolk, Virginia. Dodder seeds are frequently intercepted by port inspectors, confirming that ports are foci of introductions. Ballast and packing material have been implicated in the spread of Orobanche minor. Another source of contamination is nursery stock and the distribution of improved cultivated varieties of crop plants. The parasite seeds enter as contaminants of crop seed or soil of potted plants. Markets have been identified as foci for the spread of Striga (witchweed) seed to farmers in West Africa via contaminated crop seeds. In areas where improved crops have not been introduced, the likelihood of parasites being spread is lower. For example in Guinea, where subsistence grains contain little exotic germplasm, Striga is a limited problem.

While the spread of some root parasites and Cuscuta by humans is well documented, there are few examples of such spread for a mistletoe. One exception is Viscum album (European mistletoe) that was introduced in California as a potential crop for sale at Christmas around 1900 by Luther Burbank. An evaluation of its subsequent spread shows that it continues to expand its range at a rate of 0.35 km (0.22 miles) per year. It is interesting that the dwarf mistletoes (Arceuthobium) that are serious pathogens on conifers in the western United States have not spread to any pine species

in the southeastern U.S. This probably stems from a combination of historical factors, climate, and
host resistance.

Occasionally, indigenous plant parasites become serious pests when host species are planted in
their habitat. There are two examples in the American South: Seymeria cassioides (Figure 10), an
annual root parasite in Orobanchaceae, andPyrularia pubera , a shrub in Cervantesiaceae (formerly Santalaceae) that is endemic to the Appalachians. Foresters were first made aware of Seymeria
in the late 1960s when slash pine (Pinus elliottii) was damaged in southern Georgia. It appears that
problems with S. cassioides are exacerbated by drought conditions. Conifers grown as Christmas
trees in the mountains of West Virginia have been attacked by P. pubera, thus reducing the value
of the trees.

Of the approximately 1800 alien plants in California, only a few are parasites. The European mistletoe was mentioned earlier. Others include members of Orobanchaceae such as Phelipanche ramosa, a serious parasite of tomato and other crops, and two benign hemiparasites, Parentucel-
lia latifolia and P. viscosa. Parasitic plants apparently spread no more quickly than non-parasitic
plants. Despite this, the USDA APHIS PPQ requires permits to import or move any parasitic plant
across state lines. The Federal noxious weed list contains five genera of parasitic plants that are
considered to be the most serious pathogens: Aeginetia (not yet known in U.S.), Alectra (introduced to Puerto Rico), Cuscuta (both introduced and native species throughout the U.S.), Orobanche (including Phelipanche, introduced throughout the U.S.), and Striga (introduced to the
Carolinas as well as Florida).

Cassytha, Laurel Dodder (Lauraceae-Laurel Family)

Cassytha is a high climbing, parasitic vine with green or orange stems that bears a remarkable resemblance to Cuscuta (dodder). Cassytha is regularly mistaken for dodder, even by experienced plant scientists. The similarity between these two unrelated genera is one of the most striking parallelisms among the flowering plants. Additional features common to both Cuscuta and Cassytha are hard seeds that require scarification (breaking or scratching of the seed coat, but no host stimulant) and simultaneous attachment of a single parasite to several diverse hosts.

Cassytha is the only parasitic member of the otherwise autotrophic Laurel family. The host range of Cassytha is broad. Considering its vigor and wide distribution, it is remarkable that it is only occasionally reported as a significant pest usually on tree crops including mango (Mangifera indica) and avocado (Persea americana). Most species occur in Australia, but C. filiformis is pantropical and is also abundant in southern Florida. Because the species are superficially similar, a new introduced species could easily be overlooked.

Cuscuta, Dodder (Convolvulaceae-Morning Glory Family)

Species of Cuscuta are among the best known of all parasitic plants. The biology and control of dodders was reviewed in Dawson et al. Cuscuta has a broad host range, although monocots are less preferred. The genus Cuscuta contains three subgenera. Members of the subgenus Monogynella are robust vines that may attack and kill fruit trees, while species in the subgenus Cuscuta are more delicate in structure and favor herbaceous hosts, as do species of the entirely New World subgenus, Grammica.

Dodders may be the most important parasitic weeds of legumes in temperate regions. Of particular importance is C. campestris on alfalfa (Medicago sativa) with especially significant impact on alfalfa grown for seed. The alfalfa and dodder seeds are similar in size, and so the parasite is spread with the host. The wide range of hosts attacked by dodders is reviewed in Dawson et al. The most effective means of control is seed sanitation. Because the surface of dodder seeds is minutely roughened, dodder seeds stick to felt rollers while alfalfa seeds pass over. Dawson et al. also reviewed several herbicide treatments that are directed at the newly germinated seeds of dodder.

Orobanchaceae (Broomrape Family)

This family includes the largest number of genera (over 90) and species of all the families of parasitic flowering plants. In the past, the hemiparasitic members of this family were classified as part of Scrophulariaceae (the figwort family) while the holoparasitic members were included in Orobanchaceae (the broomrape family). Molecular phylogenetic studies have shown that all of these parasitic plants are monophyletic (i.e. all share a common ancestor) and are thus classified in Orobanchaceae. The two most economically important genera, Striga (witchweed) and Orobanche (broomrape), are discussed below, with particular attention paid to their life cycle similarities and differences.

Striga (Witchweeds)

Witchweeds (Striga spp.) have a greater impact on humans worldwide than any other parasitic plant because their hosts are subsistence crops grown widely in Africa and Asia. Such crops include maize, sorghum, pearl millet, finger millet, rice, and sugar cane as well as legume crops such as cowpea and ground nut. The name "witchweed" derives from the effect these parasites have on their host in which most damage occurs before the parasite is visible above ground.

Striga is an obligate hemiparasite that reaches its greatest diversity in the grasslands of Africa, although it also occurs in India, eastern Asia, and Australia. Two species, S. asiatica and S. hermonthica , cause the most damage to crops worldwide with often devastating yield reduction in the African Sahel. Striga species have complex life cycles involving several discrete steps: seed dispersal, after ripening, seed conditioning, germination, haustorial induction, attachment, penetration, seedling development, emergence, flowering, and fruiting. The following summary is based on reviews in Sauerborn, Parker and Riches, and Mohamed.

Striga produces thousands of "dust" seeds per capsule that are shed when the rains stop at the end of the growing season. Movement of seeds in nature is usually attributed to wind and rain, facilitated by the distinct sculpturing on the seed. Seeds are dormant for several months before they will respond to chemicals exuded by the host; this period is referred to as "after ripening." Conditioning occurs when two environmental factors are present: suitable soil temperatures in the range of 25-35 °C (77-95 °F) and moisture content near 100%. During conditioning, the seed imbibes water, and after conditioning, it is able to respond to chemical signals from the host roots. The signals indicate both the type of host and the distance to the host. Recent work has shown that host roots exude plant growth regulators called strigolactones, a group of sesquiterpene lactones.

These molecules are also known to stimulate mycorrhizal fungi colonization as well as regulate shoot and root branch formation in nonparasitic plants. Orobanchaceae have evolved the ability to recognize these chemicals as germination stimulants. Little is known about after-ripening though the role of reduction of abscisic acid levels has been suggested. Perhaps after-ripening leads to a down regulation of endogenous strigolactones making the witchweed seeds sensitive to external, host-derived strigolactones. Germination in obligate root parasites is usually cryptocotylar, that is, the cotyledons remain within the seed coat when the delicate radicle emerges. The subterranean communication between host and parasite has been intensely studied in the past decade. These complex processes are a kind of botanical "eavesdropping" with intricate molecular pirouetting. After germination, the radicle of the parasite stimulates host peroxidases and laccases that oxidize the lignin of host cell walls producing 2, 4-dimethoxybenzoquinone (DMBQ). Through redox cycling of DMBQ, signaling occurs that initiate's haustorium development on the parasite radicle. At this stage, the Striga radicle tip begins to produce structures that resemble root hairs that extend to contact the host root. These hairs "glue" the Striga root to the host. Similar hairs are produced by some other members of Orobanchaceae. If the host is suitable, the haustorium penetrates and forms a link with the host vascular system; this is the penetration stage.

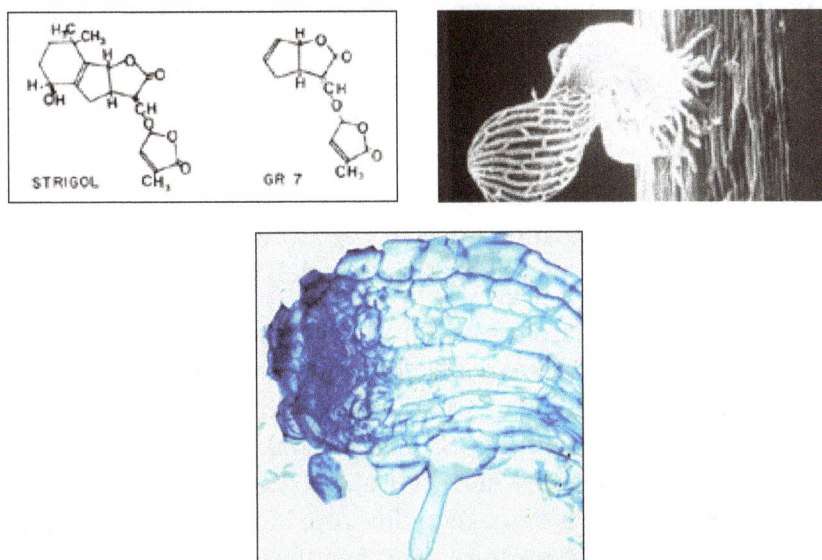

As the parasite becomes established and matures, the distinctive seedling of Striga is formed. It is subterranean, lacks chlorophyll, possesses scale leaves and produces abundant adventitious roots from which additional (lateral) haustoria can arise. The seedling exerts a powerful influence on the

growth-regulatory systems of the host by altering hormone balances and stimulating root production. Significant host damage can occur at this stage. Parasite shoots draw on host resources and
elongate upwards through the soil to reach the surface. After emergence, chlorophyll develops,
flowers form, and the life cycle is completed when the seeds are produced.

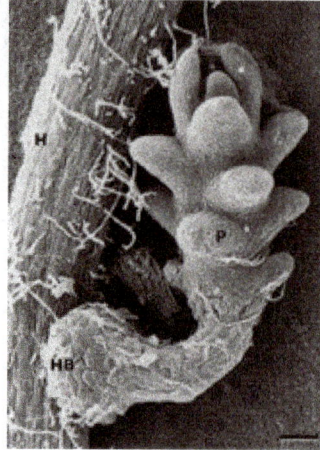

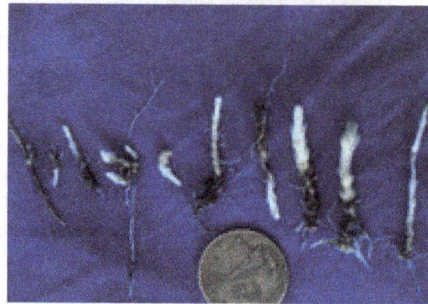

Why have so few Striga species become serious pests? One hypothesis is that the widespread S. hermonthica and S. asiatica are true agrestals, that is, plants associated only with agroecosystems. They
suggest that these crop pathogens are spread with their hosts, are more abundant in crops than
native grasslands, and have evolved from native species. Two observations support this. First, S.
hermonthica and S. asiatica have never been reported in native grasslands in Africa. Second, in areas
where there has been little introduction of improved grain varieties, Striga is a limited problem, as
noted earlier for Guinea. A similar situation exists in Togo where S. asiatica became a serious problem only after improved varieties of maize began to be widely grown.

When Striga was identified in the Carolinas, USA in 1956, federal and state action was initiated
in response to this threat. Dedicated research facilities were established to conduct basic and applied research to gain an understanding of witchweed and to develop methodology to eradicate
the infestation. Early research revealed that S. asiatica could be controlled in maize by using a
non-volatile form of the selective phenoxy herbicide 2,4-D. This method stopped the reproduction
of the Striga, but did not prevent early damage to the crop. Motorized high clearance sprayers and
backpack sprayers were used, but the magnitude of the infestation (thousands of hectares) called
for more effective methods. Moreover, the method could not be used to control witchweed growing
on weedy grasses in broadleaf crops such as cotton, soybeans and vegetables because these plants
are susceptible to this herbicide.

A major problem in witchweed control is the persistence of the microscopic seeds in the soil. Research was conducted to try to identify the seed germination stimulant. A breakthrough occurred when it was found that ethylene gas could induce Striga seed germination. New equipment and application methodologies to introduce ethylene gas into the soil were developed. Under proper soil conditions, the injection of 1.5 kilograms per hectare of ethylene stimulates the suicidal germination of up to 90% of the viable Striga seed bank in a growing season. The witchweed control and containment program in North and South Carolina is likely the most successful example of controlling of a parasitic weed. From its discovery in 1956, Striga reached its greatest distribution in 1979 when 380,000 acres were infested. By 2015 only 1,100 infested acres remained; however, it still remains a potential threat to agriculture in states as far north as Pennsylvania and west to Iowa and much of Texas, thus continued monitoring is required.

Orobanche (Broomrapes-Broomrape Family)

The genus Orobanche, as currently classified, contains ca. 78 species of obligate root holoparasites from the Old World. It is sometimes split into two genera, Orobanche and Phelipanche based on technical characters. Species of both genera are known by the English name broomrape because they were thought to grow as tubers ("rapum") from brooms (the common name for the legume Cytisus). The genus reaches its greatest diversity in Mediterranean climates and in Western Asia. All of the economically important pathogens are Old World species. New World species, previously classified as Orobanche such as O. fasciculata and O. uniflora, have been shown via molecular phylogenetic studies to belong to the genus Aphyllon. Moreover, some Old World species have also been segregated from Orobanche (genera Boulardia,Myzorrhiza, and Phelipanche).

Major crop hosts for Orobanche are legumes, solanaceous crops (eggplant, tomato, tobacco, potato but not Capsicum peppers), umbels (carrot, parsley, celery), cole crops (cabbage, cauliflower), lettuce, and sunflower. Control is difficult because of seed longevity in the soil (more than five decades), small seed size (less than the width of a human hair), fecundity (thousands of seeds per plant), and a subterranean phase (seeds germinate beneath the soil and parasitize the host before they emerge and becoming evident). Broomrapes have their greatest impact in the Balkans, the Nile Valley, Central Asia, southern India, and Nepal. Damage varies with level of infestation, and total crop failures have occurred in some cases.

There have been numerous studies of the host range of Orobanche species. It has been shown that O. ramosa (=Phelipanche ramosa) can parasitize plants from 11 different dicot families, in fact, more different hosts than any other broomrape. Major agronomically important hosts include solanaceous crops, cabbage, cauliflower, hemp, carrots, lettuce, and some legumes. The related species O. aegyptiaca (= Phelipanche aegyptiaca) causes especially severe damage to melons in

Central Asia where broomrape not only reduces the yield and weakens the melons, but also induces the production of a toxin within the melons that renders them unmarketable.

Evaluation of host range for parasitic plants is complicated by the fact that crops parasitized in pots may not be parasitized in the field. For example, there are no reports of field parasitism of soybean by any broomrapes, although soybean was parasitized by O. crenata in pots. Nonetheless, pot experiments are valuable because potential hosts can be identified and host specific strains of the parasites can be determined.

Germination requirements for Orobanche are different from those of Striga. In general, Orobanche is a parasite of colder climates, so germination temperatures range from 10 to 20 °C (50 to 68 °F), alternating with temperatures as low as 5 °C (41 °F). This may explain why O. ramosa is a problem in the Nile Valley of Sudan only in the winter, and may also explain why O. cernua attacks tobacco in the winter in India, but is not a problem on sunflower (Helianthus annuus) grown in the same region in the summer. Unlike Striga, Orobanche seedlings produce roots that are geotropically neutral, that is, they do not grow downward in response to gravity. Orobanche also differs from Striga in the way the host root system is affected. Orobanche noticeably weakens the roots of its host. An exception is O. cernua, which often forms a large, single haustorium on a strengthened host root. The role of ethylene, important for witchweed germination and a useful control for that parasite, is unclear in Orobanche germination.

The longest documented viability of Orobanche seeds in the soil was a case in Bulgaria. In 1956 some of the tobacco fields near the Tobacco Research Institute near Plovdiv were replanted in grapes due to severe broomrape infestation. In 1991 the vineyards were removed and tobacco replanted. Large numbers of O. ramosa emerged, presumably from seeds that had been dormant since the last time tobacco had been grown.

Orobanche ramosa has been introduced to California where it has been a problem on tomatoes. Despite a concerted effort to eradicate it, the parasite still persists. It is likely that seeds of the parasite were introduced with the crop, either through contaminated plants or contaminated tomato seeds. More recently it has been found at numerous sites in eastern Virginia. The relatively recent infestations of tomato in Chile should serve as a warning that we can expect O. ramosa to establish itself wherever suitable hosts are found.

Orobanche cernua is widespread in Eastern Europe and the Middle East with heavy infestations in southern India and scattered occurrences in North Africa, China, and southern Europe. Its primary hosts are solanaceous crops and sunflower. The parasite of sunflower is taxonomically classified as O. cernua var. cumana or O. cumana. Sunflower is the most important oil crop in parts of eastern Europe, and O. cernua is a major constraint on production, especially in Bulgaria where sunflower oil is very important. Infected plants are stunted and have smaller heads with aborted fruits and poorer oil quality. Between 1947 and 1950, the broomrape problem became so severe in Bulgaria that it threatened the continued cultivation of sunflower.

New, resistant varieties were introduced from the Soviet Union, but host resistance apparently selected for virulence in broomrape, with a resultant rapid loss of the resistance. Nine "races" of the parasite were documented in Bulgaria with new races recently reported from Spain. Tobacco can also be severely damaged by O. cernua. However, tobacco is not attacked in Bulgaria by O. cernua, even in fields adjacent to heavily infested sunflower fields. The situation is reversed in India, where tobacco is severely damaged and sunflower has not been attacked. This difference in host preference, studied for years by European and Russian plant breeders, supports the concept that there is genetic differentiation that might be recognized taxonomically (i.e. O. cernua vs. O. cumana). These data should be re-examined from a modern genetic perspective. More awareness of this pathogen is needed, considering the acreage devoted to sunflower in the United States.

The last of the major broomrapes to consider is O. crenata. Major hosts include broadbeans, lentils, forage legumes, carrots, parsley, and celery. It is restricted to Europe, the Middle East and North Africa. A report of O. crenata in Finland attests to the cold hardiness of this species, which has the northernmost range of the agronomically important broomrapes. Damage is severe and commodities from infected areas should be carefully monitored.

Three other genera of Orobanchaceae, Aeginetia Alectra, and Christisonia, may be considered minor problems on monocots, unusual hosts for most broomrapes. Aeginetia indica can be damaging on cereals and there are scattered reports of Christisonia scortechinii damaging sugarcane in the Philippines. Alectra is an increasing problem on cowpea and peanut (groundnut) in central Africa.

Sandalwood Parasites

The Sandalwood order (Santalales) has the widest range of trophic modes among all parasitic plants, including autotrophs, root hemiparasites, stem hemiparasites, and root holoparasites. The mistletoe habit has evolved independently five times in this order, thus mistletoe refers to the plant habit (stem parasite), not a taxonomic group. The order contains 12 genera (57 species) of autotrophs, 150 genera (2370 species) of hemiparasites, and 17 genera (42 species) of holoparasites. These plants are distributed worldwide with greatest diversity in the tropics. In terms of impacting human activity, four families deserve mention: Loranthaceae, Thesiaceae, Cervantesiaceae, and Viscaceae.

Loranthaceae (Showy Mistletoe Family)

This large, mainly tropical, family is composed of 76 genera and over 1000 species. The flowers are variable in size and shape, but many members have large showy flowers that are bird-pollinated. Indeed, the co-evolutionary relationship with birds has reached a high level in this family, as evidenced also by the seed dispersal mechanism. In Australia, birds of the genus Dicaeum have tongues that are specifically modified to sip nectar from mistletoe flowers and digestive canals that pass the viscid seeds in a remarkably short period of time. Many other mistletoe-animal interactions occur, and Watson has proposed that mistletoes function as keystone resources in many ecosystems, i.e. they are important ecological components that positively affect diversity in these habitats.

At least 30 genera of mistletoes in the Loranthaceae occur on introduced or cultivated trees, and the following have been reported to be particularly damaging: Tapinanthus bangwensis on cacao (Theobroma cacao) in Africa,Dendrophthoe pentandra on kapok (Ceiba pentandra) in Java, Passovia pyrifolia on teak (Tectona grandis) in Trinidad, and Oryctanthus occidentalis. on cacao in Costa Rica. Four genera of Loranthaceae (Agelanthus,Englerina, Globimetula, and Tapinanthus) are particularly damaging to the shea butter or karité tree (Vitellaria paradoxa, Sapotaceae) in Burkina Faso. Of the 16,000 trees examined in one study, 95% were parasitized by one or more of these mistletoes. Management of loranthaceous mistletoes generally involves pruning the mistletoe and/or the host branch.

A discussion of a typical member of Loranthaceae ("loranth" here) follows, beginning with flowering. Birds pollinate the large-flowered Old and New World species, whereas insects are the pollen vectors for small-flowered species. Some loranths have bisexual flowers, whereas others are dioecious or monoecious with unisexual flowers. Loranthaceae are unusual among angiosperms in having an "aggressive" embryo sac which actually grows out of the ovule into the ovary and even into the style in some species. After fertilization, the embryo and seed begin to form within the inferior ovary. The fruit that develops is a single-seeded berry (or drupe, but without a stony endocarp), and the enclosed seed is surrounded by sticky viscin. The viscin, composed of cellulosic strands surrounded by mucilaginous pectic material, attaches the seed to the host plant after dispersal. Unlike Orobanchaceae, loranth seeds do not require a host germination stimulant and will germinate spontaneously; however, establishment only occurs on living hosts. Emerging from the photosynthetic endosperm, the seedling radicle (or hypocotyl) is negatively phototropic and thus grows towards a dark surface (often the host branch).

The first attachment structure formed is called a holdfast, and the cotyledons either remain within the endosperm as absorptive structures (e.g. Loranthus) or emerge and expand (e.g. Psittacanthus). The haustorium forms from the holdfast, eventually connecting to the host xylem. The first aerial shoots of the mistletoe usually form from the epicotyl and, in some species, epicortical roots extend from the haustorium along the host branch. These epicortical roots form new haustoria and shoots, thus allowing lateral spread within the host branches. When in contact with the host cambium, the loranth haustorium induces the formation of additional wood that enlarges in fluted columns, eventually forming a placenta-like saddle. This structure is called a woodrose. When the mistletoe haustorial tissue (composed mainly of parenchyma) decays, the woodrose remains and is often used for decorative art objects.

Sandalwood Family Relatives

The root hemiparasitic Santalaceae that includes the sandalwood of commerce (Santalum spp.) traditionally included nearly 40 genera. Molecular phylogenetic studies resulted in a reclassification that recognizes additional families, thus reducing Santalaceae to 11 genera. Although this newly circumscribed family contains three genera of mistletoes, none are economically significant from a plant pathology perspective. Indeed, the opposite is true for Santalum album, which is extensively cultivated because it is the source for santalol, a compound used in the preparation of perfumes, cosmetics, and medicine. Two families that have pathogenic members are Thesiaceae and Cervantesiaceae. Although Thesium contains weedy species, only a few cause significant damage. In Cervantesiaceae, Pyrularia pubera can be a pathogen of trees under some circumstances in the eastern U.S. (as mentioned above). It is of interest that Okoubaka aubrevillei of West Africa, also a member of this family, can inflict severe damage on surrounding vegetation.

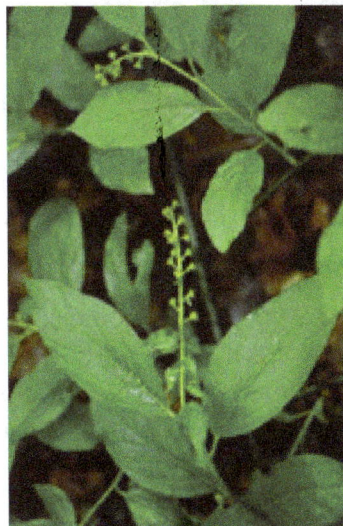

Viscaceae (Christmas Mistletoe Family)

The mistletoe family Viscaceae is small in terms of the number of genera (seven), but contains a large number of species (578). Most of these species occur in two genera: Viscum (of the Old World) and Phoradendron (of the New World). The name Viscaceae derives from a feature of the seeds - the viscin - that forms a sticky layer on its outer surface, which attaches the seeds to the host branch. These are the mistletoes most often recognized by people from temperate parts of the world because the green leafy shoots with white berries often festoon doorways around Christmas time. For more about the folklore associated with mistletoe.

Mistletoes in Viscaceae impact both positively and negatively on human activities. In addition to Christmas decoration, Viscum album is used medicinally, for example to treat various forms of cancer. Although the efficacy of some of these practices is questionable, there is growing scientific evidence of therapeutic activity. Notably, recombinant mistletoe lectin (rML) has been used to treat ovarian cancer. The other major compounds extracted from Viscum are the thionins, termed viscotoxins (VT) that not only have immunomodulatory effects, but are also strong cytotoxins. These cytotoxins are present in mistletoe berries, and thus pose a safety risk for small children who may ingest them.

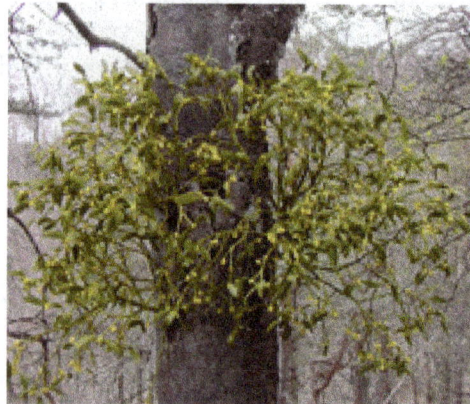

Certainly the greatest economic impact on human activity is caused by mistletoes in the genus Arceuthobium. Termed "dwarf mistletoes" because some species are diminutive, their damaging effects upon commercially important forest trees in North America are enormous: 11.3 million cubic meters of wood is lost annually. Although "leafy mistletoe" such as Phoradendron and Viscum can also cause damage to host trees, particularly in urban environments, it is not on the same scale as Arceuthobium and commercial species are seldom impacted. The leafy mistletoes occur on both hardwood and coniferous hosts, whereas Arceuthobium is known only from conifer hosts (families Pinaceae and Cupressaceae).

The life cycle of viscaceous mistletoes is similar to that described above for Loranthaceae, but with some exceptions. Pollination is generally effected by insects and wind, and the flowers in this family are very small. The monoecious or dioecious plants bear unisexual flowers on spikes or cymes. In several genera, leaves are reduced to scales. The haustorium of Viscaceae never forms epicortical roots, but instead forms a complex internal organ called the endophyte. This structure is composed of portions that run parallel to the host branch axis within the cortex, hence they are called cortical strands. Other tissues called sinkers descend perpendicularly from the cortical strands into the host xylem. Mistletoes in Viscaceae are water hemiparasites, hence they manufacture at least some of their own food through photosynthesis. Indeed, it has been documented that leafy mistletoes such as Phoradendron actually translocate photosynthates back into the host during the winter when host leaves are absent. In contrast, Arceuthobium (dwarf mistletoe) produces very little of its own food as an adult plant.

Complex types of haustoria are found in Arceuthobium. In some cases the mistletoe haustorium is localized and the only effect on the host is a smaller branch distal to the swollen infection. In other cases, a structure called a witches' broom is induced by the mistletoe haustorium that is composed of a dense group of host branches. Two types of witches' brooms are known: nonsystemic where the mistletoe endophyte does not enter the witches' broom; and systemic, where the endophyte is present, sometimes even in the host branch apical meristem. When the host cells divide by mitosis, the parasite endophyte cells also divide in a coordinated fashion. Systemic witches' brooms, as seen in A. douglasii and A. pusillum, are remarkable examples of the extremely intimate cellular relationship that has evolved between host and parasite.

In Viscum, Phoradendron, and most of the other genera of Viscaceae, seed dispersal is mediated by birds. In all but one species of Arceuthobium, birds are not the primary seed vectors. Instead the fruits have an explosive dehiscence mechanism for dispersal. The fruit walls are very elastic, and turgor pressure develops as the fruit ripens. A layer of cells that connects the fruit to the pedicel begins to weaken and, with the slightest disturbance, the fruit separates forcefully from the pedicel. This hydrostatic mechanism explosively expels the single seed at a rate of 27 meters (yards) per second, and the seed can travel up to 16 meters (yards) before sticking to a host surface. Moisture from dew or rain causes the viscin on the seed to swell, and the seed may then slide to the base of the needle fascicle where germination ensues. Although many seeds are wasted in this process, one plant may produce hundreds of fruits and seeds, thus many are successfully established on young tree branches. The seeds can only penetrate younger host tissue because there the bark is thin.

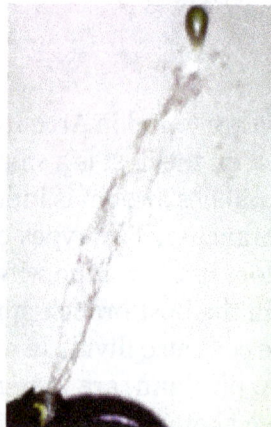

As with loranths, the seedling radicle forms a holdfast when it encounters an obstruction on the host branch, and the haustorium penetrates the host from within the holdfast. Unlike most Loranthaceae and Viscaceae, the shoots of Arceuthobium (dwarf mistletoes) are not formed from an epicotyl; once the endophyte is inside the host, the radicle (hypocotyl), cotyledons, and endosperm wither and die. The endophyte then develops within the host and aerial shoots form as much as a year later adventitiously from the cortical strands. In nature, various dwarf mistletoe species may require from three to ten years to go from seed to seed.

Controlling Dwarf Mistletoes - A Continuing Challenge

Dwarf mistletoes are often a serious threat in extensively managed forests, hence, control measures must be economical. Although there is no relationship between presence of dwarf mistletoe and site quality, once dwarf mistletoe invades a site, infected trees growing on a poor site are invariably more seriously affected than are infected trees growing on a good site. Thus, an effective form of control is to determine site quality before dwarf mistletoe has become a problem. On good sites, selective removal of infected trees during pre-commercial and commercial thinnings might be effective enough to reduce almost all losses. On poor sites, heavy dwarf mistletoe infestation might dictate early harvest or even abandonment of the stand for commercial purposes. In high-value stands such as seed orchards or recreation areas, severely infected trees should be felled. For valuable trees, pruning of infected branches can be effective if more than half the crown will remain after the infected branches are removed. The USDA Forest Service has developed a computer program that is used by forest managers to reduce losses due to dwarf mistletoe.

Natural wildfires effectively eradicated local infestations of dwarf mistletoe in the prehistoric past, and so prescribed fire has been investigated as a possible control agent. Fire has been most effective with Arceuthobium pusillum in the eastern spruce forests because this species spreads locally to form discrete infection centers with a distinct boundary between diseased and healthy trees. The use of fire has been less effective in western North American forests because here dwarf mistletoes do not generally form discrete infection centers. Instead, economic losses attributed to dwarf mistletoes are effectively minimized with silvicultural treatments that recognize stand site quality and then reduce the incidence of infected trees by thinning. In some situations, resistant tree species can be planted to replace a susceptible species presently on the site. For example, ponderosa pine can be replaced by Douglas fir, or by hardwoods such as aspen or cottonwood.

Chemical control through the use of herbicides has limited success, mainly because it is expensive and difficult to find chemical agents that affect only the mistletoe and not the host. An environmentally safe chemical, ethephon, effectively controls dwarf mistletoe in some situations. When applied in summer, the dwarf mistletoe shoots will dry up and fall off before seeds can be produced. The chemical will not affect the endophytic system, so the infected trees will need to be resprayed every few years when new dwarf mistletoe shoots appear. This technique is only useful where high-value trees are involved.

Various means of biological control have been investigated, including insects and fungi. These organisms, however, have coevolved with Arceuthobium, hence it is unlikely they can be used to dramatically reduce the size of mistletoe populations. Arceuthobium is native to North America, Asia, Europe, and Africa - the only mistletoe that occurs naturally in both the Old and New Worlds. These plants have been coevolving with their hosts for millions of years, so it is probably unrealistic to achieve total elimination. Moreover, dwarf mistletoe witches' brooms provide roosts and nesting sites for many birds and animals (including the Northern Spotted Owl), thus there is a growing tendency to take an enlightened view of the role these parasites play in the complex forest ecosystem, and to modify management practices with this view in mind.

Dodder

Dodder is a group of ectoparasitic plants with about 150 species in a single genus, Cuscuta, in the morning glory family (Convolvulaceae) or Cuscutaceae, depending on the classification system used. These pale green or yellow, bright orange or red plants, found throughout tropical and temperate regions of the world, have been given all sorts of common names alluding to the string-like appearance of the plants: angel hair, devil's hair, devil's ringlet, goldthread, hair weed, lady's laces, strangle weed, witch's hair and many others.

Orange strands of dodder nearly cover a native burro bush, Ambrosia dumosa,
in Anza Borrego Desert State Park, California.

These plants have very thin, sting-like twining stems that appear to be leafless. They do have leaves, but they are reduced to tiny scales that are barely visible. Most species have very low levels of chlorophyll, so are not green. Because of the lack of chlorophyll most produce very little food on their own, and therefore depend on their host plants for nutrition. The plants cover their host plants in a spreading, tangled mass of intertwined stems, especially where growing in full sun (twining and attachment is greatly reduced in shaded areas).

The thin, string-like stems intertwine over a plant.

Dodder species are variable in the habitat they are found in naturally and the number of different host species they can infect, with some restricted to just a few host plant species and others able to infect a wide range of hosts. A wide variety of herbaceous and small woody plants can be parasitized by dodder, including m a n y agricultural crops such as alfalfa, asparagus, carrots, cranberries, onions, and potatoes, as well as many ornamental plants, including chrysanthemum, dahlia, helenium, impatiens, English ivy, periwinkle, petunia and trumpet vine, and many weed species including field bindweed (Convolvulus arvensis), lambsquarters (Chenopodium album), and pigweed (Amaranthus species). Dodder can weaken or kill plants and reduce crops yields. The impact on the host plant varies considerably depending on the species of dodder, the growth stage and condition of the host plant, and the time of infection. Infected plants are also more susceptible to diseases and insect problems.

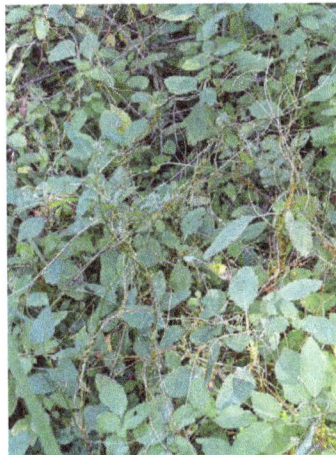

Dodder strands on jewelweed, Impatiens capensis, in Wisconsin.

In temperate areas dodder is an annual that grows from seeds each growing season. After germination at or near the soil surface, the seedlings grow quickly toward a host plant, possibly in response to volatiles released by preferred host plants, coiling around any object it encounters. If a suitable host plant is not reached within 5-10 days the seedling will not survive. Once a suitable host plant is reached, the seedling wraps itself around the plant and produces special structures, modified adventitious roots called haustoria that are inserted into the host's vascular system. Once the dodder is established on the host plant its original root in the soil dies. The dodder plant then continues to grow rapidly on that host plant, continually making new attachments to the host and eventually covering it – or spreading to adjacent plants – until the dodder plant is killed by frost. Since they have to re-establish from seedlings each year, species in temperate areas are only found in relatively low vegetation, whereas species in tropical areas, where plants can grow continuously, may reach high into the canopy of shrubs and trees.

Dodder grows on host plants, wrapping around its stems and leaves, and attaching to its vascular system.

Tiny 4 or 5-parted white, yellow or pink bell-shaped flowers are usually borne in clusters (occasionally singly) in late spring to early fall, depending on the species. These are followed by small rounded fruits or seed capsules the same color as the stems that contain 2-3 seeds. Dodder is a prolific seed producer; a single plant can produce thousands of seeds. The small seeds have a hard, rough seed coat that enables them to survive in the soil up to 20 years or more, depending on the species and environmental conditions. Dodder seeds are spread primarily through the movement of soil and equipment by humans, or in infested plant material or as a crop seed contaminant, and for some species by water.

The small flowers are produced in clusters (L and LC), with 4 or 5 petals of white or other colors, depending on the species (RC) and are followed by rounded fruits.

Dodder is not commonly found in gardens or ornamental landscapes, but if observed it should be removed immediately as dodder is difficult to control once introduced. Effective management requires a systematic approach combining several control methods over several years focused on reducing the current population, preventing seed production, and dealing with new seedlings in subsequent years. Any seedlings not yet attached to a host plant should be pulled (they are generally difficult to find, however). Once attached to a host plant, complete eradication from the host plant is usually not possible as dodder can grow back from haustoria embedded in the plant. Pruning is of little benefit unless only one or two branches are affected and can be removed without destroying or disfiguring the entire plant. So unless the plant can be pruned significantly lower than the dodder it is best to remove the entire host plant or kill both the dodder and host plant with a non-selective herbicide, such as glyphosate. If the dodder plants have already set seed, burn them (if allowed locally) or dispose of them in the trash. In areas that have been infested previously, the area should be monitored closely for the presence of seedlings and pull them as soon as they are seen or the soil can be treated with a pre-emergent herbicide before seeds germinate in spring. Control any weeds that could serve as hosts. If the area is planted annually, choose plants that are not susceptible to attack by dodder, such as ornamental grasses and other monocots, including lilies. In agricultural crops, dodder is managed by rotation to non-host crops, use of dodder-free seed, cleaning equipment between fields, control of weed hosts, and chemical control in infested areas.

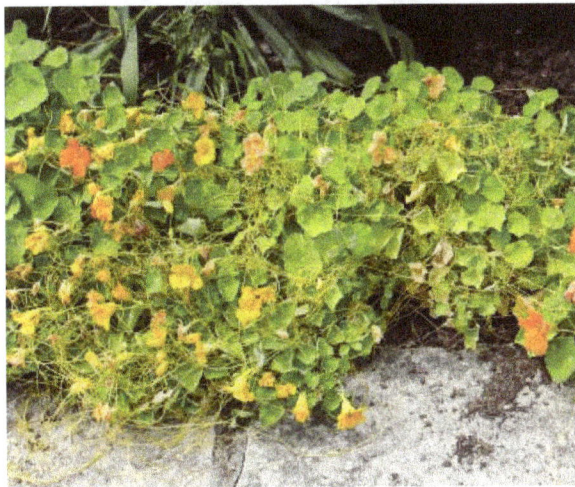

Dodder growing on nasturtium.

There are many native species of dodder in Wisconsin, including common or swamp dodder (Cuscuta gronovii (= C. umbrosa) that is potentially invasive, but not all dodder species are pests that need to be controlled. There are several species of native dodders in Wisconsin that are Wisconsin Special Concern plants:

- Hazel dodder (Cuscuta coryli) is found in sandy lake margins, mud flats and moist fields. It is often found on hazelnut (Corylus spp.), as well as reported infecting aster, common milkweed (Asclepias syriaca), Mondarda, sunflower (Helianthus sp.), Canada goldenrod (Solidago canadensis), and members of the mint family (Lamiaceae) and many other plants.

- Rope dodder (Cuscuta glomerata) is found in a variety of moist habitats, and blooms from July through September. It can be distinguished from other Cuscuta by having the inflorescences densely bunched together giving the plant a thick, rope-like appearance. It is reported as having a preference for members of the Aster family, often on sunflower, Canada goldenrod, and spotted water hemlock or spotted cowbane (Cicuta maculata).

- Field or five-angled dodder *(Cuscuta pentagona = C. arvensis, C. campestris)* is found along fields and roadsides.

- Knotweed or smartweed dodder (Cuscuta polygonorum) is also found in moist habitats including wet prairies, coastal plain marshes, and the margins of commercial cranberry bogs. It is distinguished from other Cuscuta by its very short styles and its host plants including Bidens spp., Polygonum spp., Panicum capillare, Setaria glauca, Rorippa islandica, and Hypericum boreale.

Rope dodder, Cuscuta glomerata.

Most species of dodder are on the federal noxious weed list, except some native species and a few widespread non-native species, but all dodders, including the natives, require federal permits for importation or transportation of seed.

Rhinanthus Minor – Yellow Rattle

Yellow rattle is grassland annual with yellow, two lipped flowers, the upper lip with two white or (as in our photographs) purple teeth. Behind the flower the joined sepals inflate to form a green bladder sometimes tinged with red. Flowering occurs from May through to September followed by a seed capsule which rattles within the now brown and papery bladder. The majority of seed ripens in mid to late July and the rattling of the fruit within the bladder was said to indicate that the meadow was ready to be cut for hay, hence its other common name of hay rattle.

Type	Seeds per gram	Origin	Ordering
Meadow annual	300	Somerset, Kent, Caernarvonshire	Order this species

Habitat

Yellow rattle is an annual root-hemiparasite of moderate to low fertility grasslands particularly meadows but also roadsides, dunes, lightly grazed pastures and dry areas within fens. It parasitizes the roots of a wide range of meadow plants especially grasses and legumes and whilst capable of carrying out its own photosynthesis it is dependent upon these hosts for additional supplies of carbohydrates and minerals. By drawing nutrients from surrounding vegetation it impedes their growth and helps maintain an open sward structure. Yellow rattle is however intolerant of shade and on fertile sites where coarse grasses are present or on grasslands that are never cut or grazed it tends to get shaded out. The seed, which is set from late June onward, is winged and dispersed a short distance by the wind. Seed requires a period of chilling to break dormancy and germinates in early spring. There is no persistent seed bank in the soil.

Yellow rattle populations tend to fluctuate in meadows and 'move about' from year to year as a reflection of the balance of health of the Yellow rattle plants and their host plants in any one patch.

Growing

Yellow rattle seed must be sown in autumn as it needs a prolonged period of chilling through the winter to trigger its germination the following spring. Yellow rattle can be sown as part of a mixture, or on its own into established grassland.

The most suitable sites for Yellow rattle will be grassland of low to medium fertility and with a balanced sward which is not dominated by coarse vigorous grasses. Prepare for sowing by cutting the grass very short (25mm) or by grazing very hard and open up sites for germination by harrowing, raking or lightly discing, aiming to create up to 50% bare soil. Broadcast the seed on to the prepared surface.

Traditional meadow management based around a late July hay cut provides Yellow rattle the best opportunity to set seed, and for its seed to scatter during the process of haymaking.

Sow Yellow rattle at a rate of between 0.1g and 1g per square metre. The rate chosen is most usually a compromise between the desire to rapidly attain a plant population density sufficient to suppress grass and the available budget.

Viscum Album

viscum album, also known as mistletoe is a semi-parasite, which grows mainly on deciduous trees and spread in the temperate zones of Europe and Asia. As a semi-parasite it lives on trees and cuts off the "host´s" supply of water and nutrients. The mistletoe produces energy-rich organic compounds through the photosynthesis of its own green leaves throughout the year. Thus, the host tree is not permanently damaged by the mistletoe. If the deciduous trees have shed their leaves in autumn, you can see the mistletoe on poplar, birch, willow and other deciduous trees. They are spherical subshrubs with a diameter of about 1m wide, reminiscent of a bird´s nest. Green-brown, dichotomously branched twigs spring from a short trunk. The leaves are leathery, with entire margins, lanceolate to broadly ligulate. At the end of each forked part there is a flower-bearing top. The flowers themselves are rather inconspicuous. What is remarkable are the pea-sized white berries with their slimy sticky content that form from the female flowers. Flowering time is February to May, the berries ripen in late autumn. According to an old custom the green mistletoe with the white berries are attached on the doors of the houses at Christmas time to ward off demons.

It was used to coat limed twigs and thus began the coveted meal of songbirds, especially the Mistle Thrush (Turdus viscivorus). The berries are eaten by the thrushes; the seeds pass undigested through the bird´s intestine and area disseminated in their droppings.

Applications

Segment therapy for degenerative inflammatory joint diseases by producing cuti-visceral reflexes after calming local inflammation through intradermal injections; for palliative therapy in terms of a nonspecific stimulation therapy for malignant tumours (Commission E). These applications were approved by the Commission E for mistletoe; mistletoe stems and mistletoe berries were given negative monographs (unfavourable benefit-risk ratio). The treatment monographs developed by Commission E (phytotherapy), C (anthroposophy) and D (homeopathy) on mistletoe were during the years: 1984 to 1994. Even after this time the mistletoe has been and is well researched; its immune-stimulating anti-cancer effect has been studied, focusing on the mistletoe lectins contained in the mistletoe. They influence the release of cytokines, which act as mediators for important processes of the immune system and so can also be used to fight tumour cells. Cytokines however, have two faces, because they can stimulate the division of tumour cells and thus by improving the immune status accelerates the growth of the tumour. Clinical trials with cancer patients who were treated with a standardised phytotherapeutic with mistletoe lectin content have so far have not shown any remarkable effects. However, there was improvement in the quality of life, appetite, mood and the general condition and performance of the patients. The evidence base has led to the approval of mistletoe injection preparations "for the palliative treatment in terms of a nonspecific stimulation therapy for malignant tumours" (allopathic phytotherapy). Treatment is in the hands of experienced doctors. Another approach is the background of anthroposophical cancer therapy with anthroposophical mistletoe preparations. It goes back to Rudolf Steiner, who introduced the use of mistletoe extracts in cancer therapy in 1921. He declared the special spirituality of anthroposophical mistletoe preparations as part of his metaphysical and esoteric, dogmatic pharmacology was a causal factor in healing. Such preparations are designated in pharmaceutical legislation as "Special treatment directions" and may according to the findings of the

Commission C (formerly responsible for anthroposophic medicinal products) be used "for the treatment of malignant and benign tumour diseases, malignant diseases and associated disorders of the blood-forming organs and stimulate bone marrow growth and to prevent tumour recurrence after surgery." This form of treatment can only be performed by trained doctors. The scientific evidence is pending.

Traditional use

Mistletoe is also traditionally used in combination with other drugs to support the cardiovascular function.

Medicinal Herbal Preparations in Finished Drug Products

1. Phytotherapy for the palliative treatment of tumours:

   - Aqueous extract in injection solutions,

   - Fluid extract in injection solutions.

2. Phytotherapy to support the circulatory function:

   - Cut drug to prepare tea,

   - Powder in tablets,

   - Dried extract in pills

   - Tincture and alcoholic extracts in drops,

   - Mother tincture in drops.

3. Anthroposophic medicine for cancer therapy:

   - Fresh vegetable juices from different mistletoe host trees in injection solutions,

   - Aqueous extracts from fermented apple tree mistletoe in injection solutions.

Rafflesia Arnoldii

The Rafflesia arnoldii, also known as the "corpse flower" plant, is unique in every way possible. This plant is a complete parasite lacking roots, stems, and leaves of any kind but lives attached to its host plant with flowers visible on the surface of the host plant as the only evidence of existence of Rafflesia. The main body of the plant, consisting of thread-like strands of tissue is completely embedded within the host's tissues. The flowers are produced from large buds that are about 30 centimeters wide, and the flowers themselves are as large as 3 feet in diameter. The five lobed flowers are reddish brown in color with white spots. The lobes of the flower appear to emerge from the base of a cup like structure which hosts a cylindrical column with a disc. The flowers might be either male or female with anthers or styles growing underneath the disc.

Reproduction and Life Cycle

When ready to reproduce, the Rafflesia generates maroon- or magenta-colored buds that develop over a period of a year into a large cabbage shaped size and finally blooms to form the gigantic flower. The foul, rotten flesh-type of smell of the flower attracts carrion feeding flies like those belonging to the genera Luciliaand Sarcophaga. Though the flies receive no benefits from the flower, when they sit on the flower, attracted by its smell, the pollen adhere to their backs. When these flies move to a female flower, they deposit the pollen on these flowers, allowing fertilization to occur. The fruits produced are small and fleshy with thousands of seeds. These fruits are consumed by tree shrews, which then help disperse the seeds of the plant. Since the Rafflesia is a unisexual plant and rare in occurrence, there are very rare chances that a fly that sat on a male flower and bearing pollen from that flower will sit on a female flower to transfer the pollen to the female for fertilization.

Distribution and Range

Since the Rafflesia flower lasts for only a few days, only a few lucky individuals can catch a glimpse of this flower. There are two varieties of Rafflesia arnoldii found in the wild, both endemic to Indonesia. The R. arnoldii var. arnoldi is found in the Indonesian islands of Borneo and Sumatra. The R. arnoldii var. atjehensis is found in northern Sumatra which differs from the former variety in missing a part of the ramenta in its central column.

Threats and Conservation

Currently, Rafflesia arnoldii is regarded as one of the most threatened species of plants on earth. Some species of Rafflesia, like the Rafflesia magnifica, are even classified as "critically endangered" by the International Union for the Conservation of Nature (IUCN). The small range of distribution of this species and the destruction of habitat of Rafflesia are the two primary factors driving these species to extinction. Though environmentalists have attempted to grow the Rafflesia in controlled and protected environments, such efforts have been largely unsuccessful. Some private properties in Indonesia have Rafflesia growing within boundaries of the property. The owners of such properties have been encouraged by the government to save the flowers and exhibit them to the public by charging fees for the sighting.

References

- Parasitic-plant, plant: britannica.com, Retrieved 3 March, 2019
- ParasiticPlants: apsnet.org, Retrieved 13 January, 2019
- Dodder: wimastergardener.org, Retrieved 14 May, 2019
- Mistletoe, medicinal-plants: koop-phyto.org, Retrieved 24 July, 2019
- Rafflesia-arnoldii-the-largest-flower-on-earth: worldatlas.com, Retrieved 10 February, 2019

Chapter 4
Helminths and Arthropod Parasites

Helminths are a type of parasitic worms which feed on a living host in order to survive. Round-worms, tapeworms and hookworms are some of the different types of helminths. Arthropods are a part of numerous parasitic relationships, as both parasites and hosts. This chapter has been carefully written to provide an easy understanding of helminthes and arthropod parasites.

Helminths

All helminths are multicellular eukaryotic invertebrates with tube-like or flattened bodies exhibiting bilateral symmetry. They are triploblastic (with endo-, meso- and ecto-dermal tissues) but the flatworms are acoelomate (do not have body cavities) while the roundworms are pseudocoelomate (with body cavities not enclosed by mesoderm). In contrast, segmented annelids (such as earthworms) are coelomate (with body cavities enclosed by mesoderm).

Many helminths are free-living organisms in aquatic and terrestrial environments whereas others occur as parasites in most animals and some plants. Parasitic helminths are an almost universal feature of vertebrate animals; most species have worms in them somewhere.

Biodiversity

Three major assemblages of parasitic helminths are recognized: the Nemathelminthes (nematodes) and the Platyhelminthes (flatworms), the latter being subdivided into the Cestoda (tapeworms) and the Trematoda (flukes):

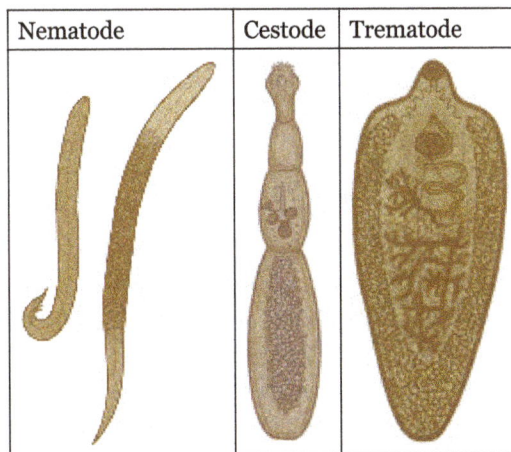

Nematode	Cestode	Trematode

- Nematodes (roundworms) have long thin unsegmented tube-like bodies with anterior mouths and longitudinal digestive tracts. They have a fluid-filled internal body cavity (pseudocoelum) which acts as a hydrostatic skeleton providing rigidity (so-called 'tubes under pressure'). Worms use longitudinal muscles to produce a sideways thrashing motion. Adult worms form separate sexes with well-developed reproductive systems.

- Cestodes (tapeworms) have long flat ribbon-like bodies with a single anterior holdfast organ (scolex) and numerous segments. They do not have a gut and all nutrients are taken up through the tegument. They do not have a body cavity (acoelomate) and are flattened to facilitate perfusion to all tissues. Segments exhibit slow body flexion produced by longitudinal and transverse muscles. All tapeworms are hermaphroditic and each segment contains both male and female organs.

- Trematodes (flukes) have small flat leaf-like bodies with oral and ventral suckers and a blind sac-like gut. They do not have a body cavity (acoelomate) and are dorsoventrally flattened with bilateral symmetry. They exhibit elaborate gliding or creeping motion over substrates using compact 3-D arrays of muscles. Most species are hermaphroditic (individuals with male and female reproductive systems) although some blood flukes form separate male and female adults.

Unlike other pathogens (viruses, bacteria, protozoa and fungi), helminths do not proliferate within their hosts. Worms grow, moult, mature and then produce offspring which are voided from the host to infect new hosts. Worm burdens in individual hosts (and often the severity of infection) are therefore dependent on intake (number of infective stages taken up). Worms develop slowly compared to other infectious pathogens so any resultant diseases are slow in onset and chronic in nature. Although most helminth infections are well tolerated by their hosts and are often asymptomatic, subclinical infections have been associated with significant loss of condition in infected hosts. Other helminths cause serious clinical diseases characterized by high morbidity and mortality. Clinical signs of infection vary considerably depending on the site and duration of infection. Larval and adult nematodes lodge, migrate or encyst within tissues resulting in obstruction, inflammation, oedema, anaemia, lesions and granuloma formation. Infections by adult cestodes are generally benign as they are not invasive, but the larval stages penetrate and encyst within tissues leading to inflammation, space-occupying lesions and organ malfunction. Adult flukes usually cause obstruction, inflammation and fibrosis in tubular organs, but the eggs of blood flukes can lodge in tissues causing extensive granulomatous reactions and hypertension.

Life-cycles

Helminths form three main life-cycle stages: eggs, larvae and adults. Adult worms infect definitive hosts (those in which sexual development occurs) whereas larval stages may be free-living or parasitize invertebrate vectors, intermediate or paratenic hosts. Nematodes produce eggs that embryonate in utero or outside the host. The emergent larvae undergo 4 metamorphoses (moults) before they mature as adult male or female worms. Cestode eggs released from gravid segments embryonate to produce 6-hooked embryos (hexacanth oncospheres) which are ingested by intermediate hosts. The oncospheres penetrate host tissues and become metacestodes (encysted larvae). When eaten by definitive hosts, they excyst and form adult tapeworms. Trematodes have more complex life-cycles where 'larval' stages undergo asexual amplification in snail intermediate hosts. Eggs hatch to release free-swimming miracidia which actively infect snails and multiply in sac-like sporocysts to produce numerous rediae. These stages mature to cercariae which are released from the snails and either actively infect new definitive hosts or form encysted metacercariae on aquatic vegetation which is eaten by definitive hosts.

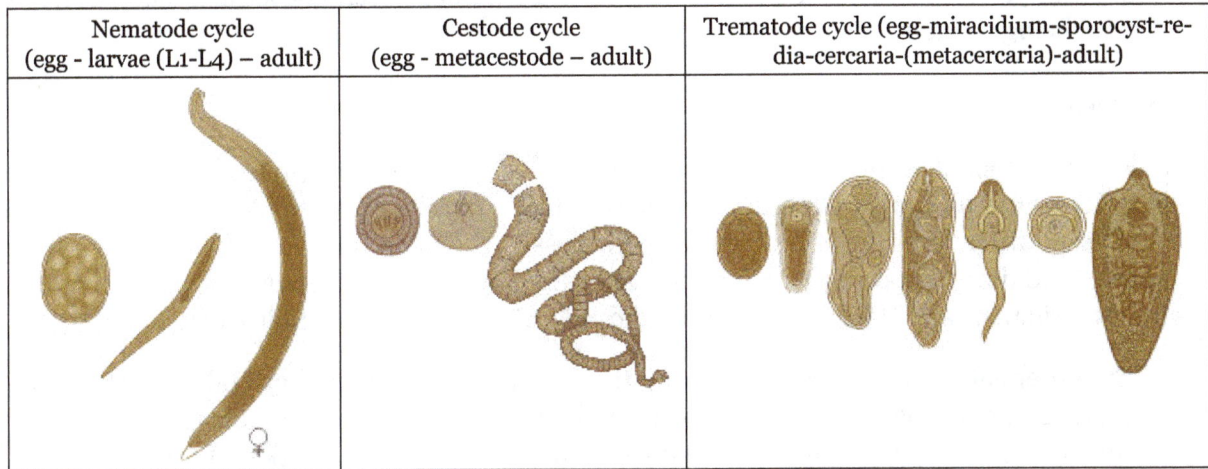

Nematode cycle (egg - larvae (L1-L4) – adult)	Cestode cycle (egg - metacestode – adult)	Trematode cycle (egg-miracidium-sporocyst-re-dia-cercaria-(metacercaria)-adult)

Helminth eggs have tough resistant walls to protect the embryo while it develops. Mature eggs hatch to release larvae either within a host or into the external environment. The four main modes of transmission by which the larvae infect new hosts are faecal-oral, transdermal, vector-borne and predator-prey transmission:

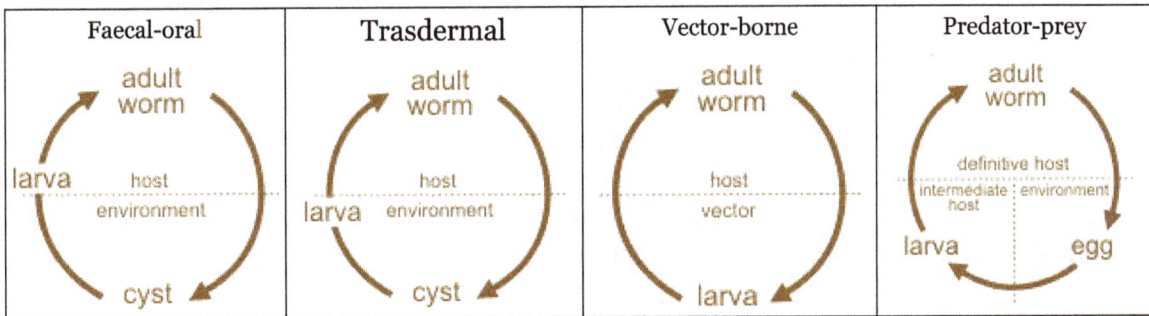

- Faecal-oral transmission of eggs or larvae passed in the faeces of one host and ingested with food/water by another (e.g. ingestion of Trichuris eggs leads directly to gut infections in humans, while the ingestion of Ascaris eggs and Strongyloides larvae leads to a pulmonary migration phase before gut infection in humans).

- Transdermal transmission of infective larvae in the soil (geo-helminths) actively penetrating the skin and migrating through the tissues to the gut where adults develop and produce eggs that are voided in host faeces (e.g. larval hookworms penetrating the skin, undergoing pulmonary migration and infecting the gut where they feed on blood causing iron-deficient anaemia in humans).

- Vector-borne transmission of larval stages taken up by blood-sucking arthropods or undergoing amplification in aquatic molluscs (e.g. Onchocerca microfilariae ingested by blackflies and injected into new human hosts, Schistosoma eggs release miracidia to infect snails where they multiply and form cercariae which are released to infect new hosts).

- Predator-prey transmission of encysted larvae within prey animals (vertebrate or invertebrate) being eaten by predators where adult worms develop and produce eggs (e.g. Dracunculus larvae in copepods ingested by humans leading to guinea worm infection, Taenia cysticerci in beef and pork being eaten by humans, Echinococcus hydatid cysts in offal being eaten by dogs).

Taxonomic

Two classes of nematodes are recognized on the basis of the presence or absence of special chemo-receptors known as phasmids: Secernentea (Phasmidea) and Adenophorea (Aphasmidea). While many different orders are recognized within these classes, the main parasitic assemblages infecting humans and domestic animals include one aphasmid order (Trichocephalida) and 6 phasmid orders (Oxyurida, Ascaridida, Strongylida, Rhabditida, Camallanida, and Spirurida).

- Trichocephalid 'whip-worms' have long thin anterior ends which they embed in the intestinal mucosa of their hosts. They have simple life-cycles where infections are acquired by the ingestion of eggs and emergent larvae moult and mature to adults in the gut. Trichuris infections in humans may cause inflammation, tenesmus, straining and rectal prolapse.

- Oxyurid 'pin-worms' have small thin bodies with blunt anterior ends. They have simple life-cycles, but with an unusual modification. Female worms emerge from the anus of their hosts at night and attach eggs to the skin. This causes peri-anal itching and eggs are transferred by hand to mouth. Infections by Enterobius cause irritability and sleeplessness in humans, especially children.

- Ascarid 'roundworms' have large bodies with 3 prominent anterior lips. Their life-cycles involve a stage of pulmonary migration where larvae released from ingested eggs invade the tissues and migrate through the lungs before returning to the gut to mature as adults. Ascaris infections in humans cause gastroenteritis, protein depletion and malnutrition and heavy infections can cause gut obstruction.

- Strongyle 'hookworms' have dorsally curved mouths armed with ventral cutting plates or teeth which they embed in host tissues to feed on blood. They have complex life-cycles where larvae develop in the external environment (as 'geo-helminths') before infecting hosts by penetrating the skin. Once inside, they undergo pulmonary migration before settling in the gut to feed. Heavy infections by Ancylostoma and Necator cause severe iron-deficient anaemia in humans, especially children.

- Rhabditid 'threadworms' have tiny bodies which become embedded in the host mucosa. Their life-cycle includes parasitic parthenogenetic females producing eggs which may hatch internally (leading to auto-infection) or externally (leading to transmission of infection or formation of free-living male and female adults). Super-infections by Strongyloides may cause severe haemorrhagic enteritis in humans.

- Camallanid 'guinea worms' infect host tissues where the large females cause painful blisters on the feet and legs. When hosts seek relief by immersion in water, the blisters rupture releasing live larvae which infect copepods that are subsequently ingested with contaminated drinking water. The 'fiery serpents' mentioned in historical texts are thought to refer to Dracunculus infections.

- Spirurid 'filarial worms' occur as long thread-like adults in blood vessels or connective tissues of their hosts. The large female worms release live larvae (microfilariae) into the blood or tissues which are taken up by blood-sucking mosquitoes or pool-feeding flies and transmitted to new hosts. Onchocerca infections cause nodules, skin lesions and blindness in humans, while those of Wuchereria cause elephantitis.

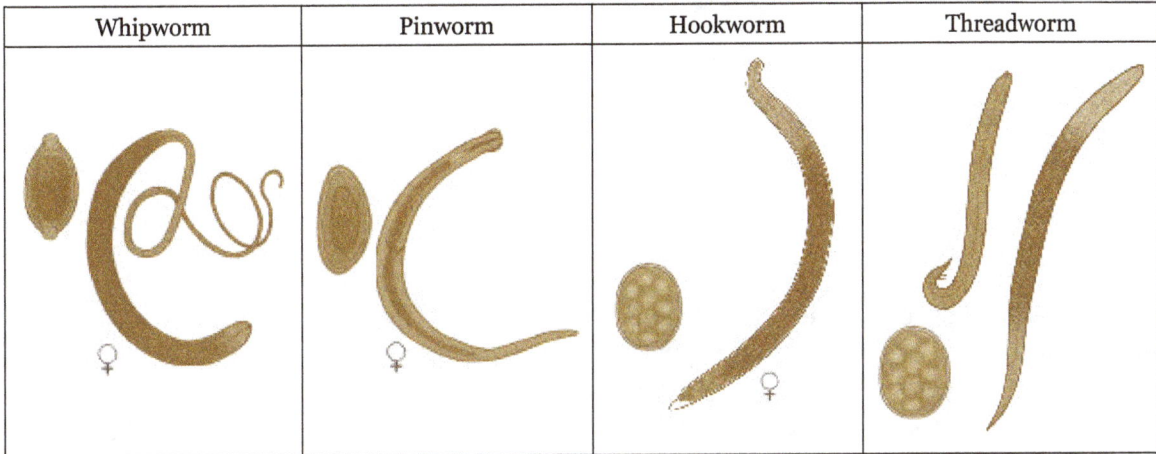

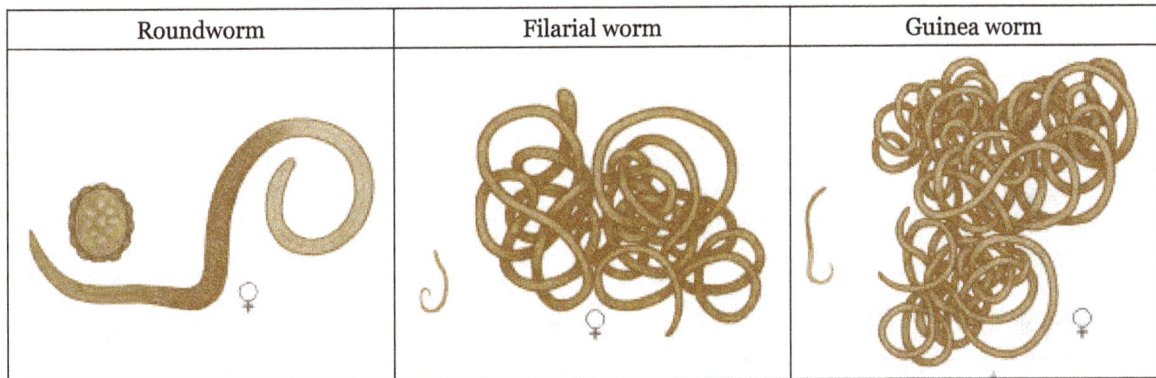

Whipworm	Pinworm	Hookworm	Threadworm

Roundworm	Filarial worm	Guinea worm

Two subclasses of cestodes are differentiated on the basis of the numbers of larval hooks, the Cestodaria being decacanth (10 hooks) and the Eucestoda being hexacanth (6 hooks). Collectively, 14 orders of cestodes have been identified according to differences in parasite morphology and developmental cycles. Two orders have particular significance as parasites of medical and veterinary importance:

- Cyclophyllidean cestodes have terrestrial 2-host life-cycles where adult tapeworms develop in carnivores (scolex with 4 suckers and sometimes hooks) while larval metacestodes form bladder-like cysts in the tissues of herbivores. The larvae of Taenia spp. cause cysticercosis in cattle, pigs and humans, while those of Echinococcus cause hydatid disease in humans, domestic and wild animals.

- Pseudophyllidean cestodes have aquatic 3-host life-cycles, involving the sequential formation of adult tapeworms in fish-eating animals (scolex with 2 longitudinal bothria), procercoid larval stages in aquatic invertebrates (copepods) and then plerocercoid (spargana) stages in fish e.g. Diphyllobothrium in humans, dogs and cats being transmitted through copepods and fish.

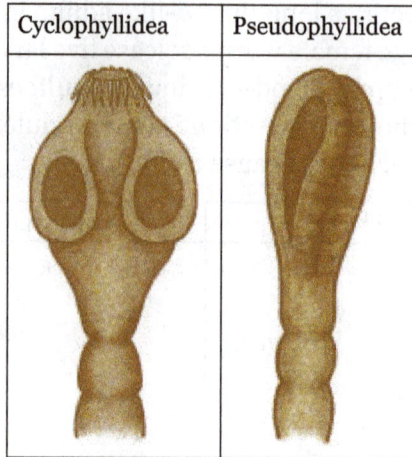

Cyclophyllidea	Pseudophyllidea

Two major groups of trematodes are recognized on the basis of their structure and development: monogenean trematodes with complex posterior adhesive organs and direct life-cycles involving larvae called oncomiracidia; and digenean trematodes with oral and posterior suckers and heteroxenous life-cycles where adult worms infect vertebrates and larval miracidia infect molluscs to proliferate and produce free-swimming cercariae. Monogenea are almost exclusively ectoparasites of fishes while Digenea are endoparasites in many vertebrate hosts and have snails as vectors. Some 10 digenean orders are recognized on the basis of morphologic and biologic differences, two orders are of particular medical and veterinary significance.

- Echinostomatid fasciolids (liver flukes) live as adults in hepatic bile ducts of mammals where they cause fibrotic 'pipestem' disease. The parasites proliferate in freshwater snails and mammals become infected by ingesting metacercariae attached to aquatic vegetation. Several Fasciola spp. cause hepatic disease in domestic ruminants and occasionally in humans.

- Strigeatid schistosomes (blood flukes) are unusual in that the adults are not hermaphroditic but form separate sexes which live conjoined in mesenteric veins in mammals. Female worms lay eggs which actively penetrate tissues to be excreted in urine/faeces or they become trapped in organs where they cause granuloma formation. Miracidia released from eggs infect aquatic snails and produce fork-tailed cerceriae which actively penetrate the skin of their hosts. Several Schistosoma spp. cause schistosomiasis/bilharzia in humans.

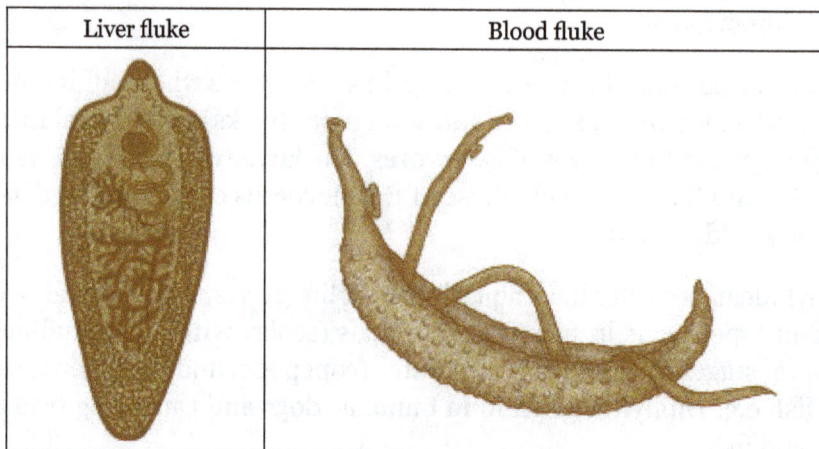

Liver fluke	Blood fluke

Parasitic Nematodes

Nematodes are small worms found in water, soil, plants and animals, and there are roughly 10,000 known species throughout the world. While some nematodes are free-living, others are parasitic and need other organisms (called hosts) to keep themselves alive. Once attached to their host, they divert nutrients and feed off of blood, tissues or pieces of cells to facilitate their own growth. While in some cases these parasitic nematodes can help control pests, in other cases they can cause damage, illness or death to the host organism.

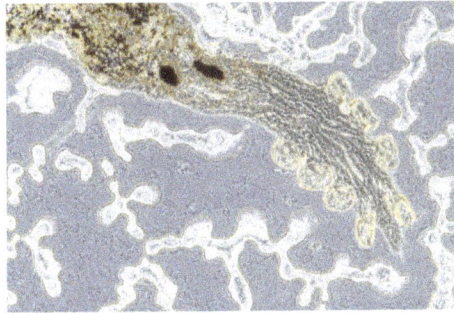

Human and Animal Parasitic Nematodes

Human parasitic nematodes like hookworm, whipworm, Ascaris (parasitic roundworm), filarial worms, eyeworms, trichinella, tapeworms and flukes are estimated to infect as many as 3 billion people. Each of these worms has adapted to its environment to find a unique way of entering its human host. Some enter our bodies by burrowing directly through the skin from soil or water, while others make their way to our intestines in the food we eat.

Many of these parasitic nematodes also infect animals, livestock and pets. For example, eyeworm attacks baboons as well as humans, and many species closely related to Ascaris infect dogs, cats, cattle, chickens, pigs and horses. In some cases, animals may be intermediate hosts where the nematodes enter and grow for a period of time as larvae and then become dormant cysts. If a human eats the infected meat, the cysts become active larvae again and grow into adult worms. This is the case with tapeworms that infect cows, fish or pigs and then latch onto the intestinal wall of the human that consumes them. Similarly, the trichinella roundworm lives and mates in the intestines of pigs, rats and other animals and, when those animals are eaten by other carnivores (humans or other animals), the parasite is passed on, causing a disease called Trichinosis. Heartworm is another common animal parasitic nematode that infects pets.

Plant Parasitic Nematodes

Plant parasitic nematodes invade the roots of plants and position themselves to divert nutrients away from the plant toward their own growth. There are two types of plant parasitic nematodes. Ectoparasites feed from the outside of plant tissue and endoparasites enter the plant tissue in order to feed. These parasites destroy the plant by damaging its vascular tissue and interfering with the transport of nutrients or by creating open wounds that leave it susceptible to other pathogens. One type of plant parasitic nematode called root knot (species Meloidogyne) causes an estimated

80 billion dollars in crop damage annually. Other plant parasitic nematode species include root-lesion nematodes (Pratylenchus), pin nematodes (Paratylenchus), ring nematodes (Criconemella), stubby-root nematodes (Trichodorus and Paratrichodorus), daggar nematodes (Xiphinema) and "mint nematodes" (Longidorus).

Insect Parasitic Nematodes

Insect parasitic nematodes are called entomopathogenic. They are free-living as adults but infect a host insect during their larval stage. They remain in the host insect until they grown to juvenile stage, and then they exit the insect by rupturing a hole in the host's cuticle. While some insects survive this exit, most die. Insect parasitic nematodes are sometimes used as biological control agents because they can be produced and used in mass numbers to attack and kill insect pests such as blackflies and mosquitoes.

Eight other families of insect parasitic nematodes contain species that attack and use insects as hosts, including Allantone-matidae, Neotylenchidae, Mermithidae, Diplogasteridae, Heterorhabditidae, Sphaerulariidae, Rhabditidae, Steinernematidae and Tetradonematidae.

Tapeworm

Tapeworm, name for the parasitic flatworms forming the class Cestoda. All tapeworms spend the adult phase of their lives as parasites in the gut of a vertebrate animal (called the primary host). Most tapeworms spend part of their life cycle in the tissues of one or more other animals (called intermediate hosts), which may be vertebrates or arthropods.

Anatomy and Function

An adult tapeworm consists of a knoblike head, or scolex, equipped with hooks for attaching to the intestinal wall of the host (which may be a human), a neck region, and a series of flat, rectangular body segments, or proglottids, generated by the neck. The chain of proglottids may reach a length of 15 or 20 ft (4.6–6.1 m). Terminal proglottids break off and are excreted in the feces of the host, but new ones are constantly formed at the anterior end of the worm. As long as the scolex and neck are intact the worm is alive and capable of growth. A rudimentary nervous system and excretory system run the length of the worm, through the proglottids. However, there is no digestive tract; the worm absorbs the host's digested food through its cuticle, or outer covering.

Reproduction

Each proglottid contains a complete set of male and female reproductive organs that produce the sex cells. Fertilization is internal; in most species cross fertilization between two adjacent worms is necessary, but in a few species self-fertilization may occur between two proglottids of the same worm, or within the same proglottid. In some species the fertilized eggs are shed continuously and leave the host's body in the feces; in others the fertilized eggs are stored until the proglottid is filled with them and the entire proglottid is then shed. The eggs develop into embryos with a hard outer shell; these do not hatch until they are eaten by a suitable intermediate host.

Humans as Tapeworm Hosts

Human tapeworm infestations are most common in regions where there is fecal contamination of soil and water and where meat and fish are eaten raw or lightly cooked. In the case of the human tapeworm most common in the United States (the beef tapeworm, *Taenia saginata*) the usual intermediate host is a cow, which ingests the proglottid while drinking or grazing. The round-bodied embryos, equipped with sharp hooks, hatch and bore through the cow's intestinal wall into the bloodstream, where they are carried to the muscles. Here each embryo encloses itself in a cyst, or bladder; at this stage it is called a bladder worm. During the bladder worm stage the embryo develops into a miniature scolex; it remains encysted until the muscle is eaten by a primary host, in this case a human. If the scolex has not been killed by sufficient cooking of the meat, it sheds its covering and attaches to the intestinal wall, where it begins producing proglottids.

A human tapeworm common in Mexico, the pork tapeworm (*T. solium*), has a similar life cycle, with a pig as the usual intermediate host. The fish tapeworm, *Diphyllobothrium latum,* transmitted to humans from fish, especially pike, is common in Asia and in Canada and the northern lake regions of the United States. This tapeworm has a more elaborate life cycle, involving both a fish and a crustacean as intermediate hosts. The dwarf tapeworm, *Hymenolepsis nana,* is transmitted through fecal contamination and is common in children in the southeastern United States. There are also several tapeworms for whom humans are the usual intermediate host; among these, the dog tapeworm, *Echinococcus granulosis,*spends its adult phase in the intestines of dogs.

Consequences of Infestation

Intestinal tapeworm infestation frequently occurs without symptoms; occasionally there is abdominal discomfort, diarrhea, constipation, or weight loss. The presence of tapeworm proglottids in clothing, bedding, or feces is the usual sign of infestation. Treatment is typically with albendazole or praziquantel, which kill the worm.

The most serious tapeworm infestation in humans is caused by the ingestion of *T. solium* eggs through fecal contamination, which results in the person serving as the intermediate, rather than the primary, host. The embryos migrate throughout the body, producing serious illness if they lodge in the central nervous system. In many poorer regions of the world, the larvae of *T. solium* are a major cause of human epilepsy. The embryos of the dog tapeworm encyst in various internal organs of humans, most commonly in the liver. The cysts produced by these embryos are called hydatid cysts, and the infestation of the liver is called hydatid disease.

Classification

Tapeworms are classified in the phylum Platyhelminthes, class Cestoda.

Platyhelminth

Flatworm, also called platyhelminth, is any of the phylum Platyhelminthes, a group of soft-bodied, usually much flattened invertebrates. A number of flatworm species are free-living, but about 80 percent of all flatworms are parasitic—i.e., living on or in another organism and securing nourishment from it. They are bilaterally symmetrical (i.e., the right and left sides are similar) and lack specialized respiratory, skeletal, and circulatory systems; no body cavity (coelom) is present. The body is not segmented; spongy connective tissue (mesenchyme) constitutes the so-called parenchyma and fills the space between organs. Flatworms are generally hermaphroditic—functional reproductive organs of both sexes occurring in one individual. Like other advanced multicellular animals, they possess three embryonic layers—endoderm, mesoderm, and ectoderm—and have a head region that contains concentrated sense organs and nervous tissue (brain). Most evidence, however, indicates that flatworms are very primitive compared with other invertebrates (such as the arthropods and annelids). Some modern evidence suggests that at least some flatworm species may be secondarily simplified from more complex ancestors.

The phylum consists of four classes: Trematoda (flukes), Cestoda (tapeworms), Turbellaria (planarians), and Monogenea. It should be noted that some authorities consider Monogenea, which contains the order Aspidogastrea, to be a subclass within the class Trematoda. Members of all classes except Turbellaria are parasitic during all or part of the life cycle. Most turbellarians are exclusively free-living forms. More than 20,000 flatworm species have been described.

General Features

Importance

Although some platyhelminths (flatworms) are free-living and nondestructive, many other species (particularly the flukes and tapeworms) parasitize humans, domestic animals, or both. In Europe, Australia, and North and South America, tapeworm infestations of humans have been greatly reduced as a consequence of routine meat inspection. But where sanitation is poor and meat eaten undercooked, the incidence of tapeworm infestations is high. In the Baltic countries much of the population is infested with the broad tapeworm (*Diphyllobothrium latum*); in parts of the southern United States a small proportion of the population may be infested with the dwarf tapeworm (*Hymenolepis nana*). In Europe and the United States the beef tapeworm (*Taenia saginata*) is common because of the habit of eating undercooked steaks or other beef products.

Parasites in immature stages (larvae) can cause serious damage to the host. A larval stage of the gid parasite of sheep (Multiceps multiceps) usually lodges in the sheep brain. Fluid-filled hydatid cysts (i.e., sacs containing many cells capable of developing into new individuals) of Echinococcus may occur almost anywhere in the body of sheep. In humans, hydatids of the liver, brain, or lung are often fatal. Infestation occurs only where people live in close association with dogs that have access to infested sheep for food.

Thirty-six or more fluke species have been reported as parasitic in humans. Endemic (local) centres of infection occur in virtually all countries, but widespread infections occur in the Far East, Africa, and tropical America. Many species are ingested as cysts, called metacercariae, in uncooked food—e.g., the lung fluke Paragonimus westermani found in crayfish and crabs, the intestinal flukes Heterophyes heterophyes and Metagonimus yokogawai and the liver fluke Opisthorchis sinensis in fish, and the intestinal fluke Fasciolopsis buski on plants. Free-swimming larvae (called cercariae) of blood flukes penetrate the human skin directly. In humans these parasites and others listed cause much misery and death. Control of certain flukes through the eradication of their mollusk hosts has been attempted but without much success.

Schistosomiasis (bilharziasis) is a major human disease caused by three species of the genus Schistosoma, known collectively as blood flukes. Africa and western Asia (e.g., Iran, Iraq) are endemic centres for S. haematobium; S. mansoni also is found in these areas, as well as in the West Indies and South America. In the Far East, S. japonicum is the important blood fluke.

Among domestic animals, the sheep liver fluke (Fasciola hepatica) may cause debilitating and fatal epidemics (liver rot) in sheep. These animals become infected by eating metacercariae encysted on grass. Monogenea are common pests on fish in hatcheries and home aquariums.

Size Range

Most turbellarians are less than five millimetres (0.2 inches) long, and many are microscopic in size. The largest of this class are the planarians, which may reach 0.5 metre (about 20 inches) in length. Trematodes are mostly between about one and 10 millimetres (0.04 to 0.4 inch) long; members of some species, however, may grow to several centimetres. The smallest cestodes are about one millimetre (0.04 inches) long, but members of a few species exceed 15 metres (50 feet) in length. The Monogenea range in length from 0.5 to 30 millimetres (0.02 to 1.2 inches). Aspidogastrea are from a few millimetres to 100 millimetres in length.

Distribution and Abundance

In general, free-living flatworms (the turbellarians) can occur wherever there is moisture. Except for the temnocephalids, flatworms are cosmopolitan in distribution. They occur in both fresh water and salt water and occasionally in moist terrestrial habitats, especially in tropical and subtropical regions. The temnocephalids, which are parasitic on freshwater crustaceans, occur primarily in Central and South America, Madagascar, New Zealand, Australia, and islands of the South Pacific.

Some flatworm species occupy a very wide range of habitats. One of the most cosmopolitan and most tolerant of different ecological conditions is the turbellarian Gyratrix hermaphroditus, which occurs in fresh water at elevations from sea level to 2,000 metres (6,500 feet) as well as in saltwater pools. Adult forms of parasitic flatworms are confined almost entirely to specific vertebrate hosts; the larval forms, however, occur in vertebrates and in invertebrates, especially in mollusks, arthropods (e.g., crabs), and annelids (e.g., marine polychaetes). They are cosmopolitan in distribution, but their occurrence is closely related to that of the intermediate host or hosts.

Roundworms

Roundworms are parasites that live in your intestine. A parasite is a creature that lives in or on another creature in order to survive. They have long round bodies and range in size. Roundworms can live in or on humans, and can cause many problems. They are usually found in soil and stool and can enter the body through the mouth or direct contact with the skin. They can live in the human intestine for a very long time. There are several types of roundworms and they can all be quite harmful.

Anyone can get roundworms. Poverty-stricken individuals living in underdeveloped areas of the world are most susceptible to roundworms. School-aged children and people who are institutionalized are also susceptible. Poor-hygiene practices are a big contributing factor to contracting roundworms. Roundworms grow best in warm to hot climates, so people in these climates need to be extra aware of the symptoms of roundworms.

Types, Causes, Transmission and Symptoms

- Ascariasis:

    ◦ How it is transmitted: Mostly transmitted through poor hygiene. It is usually found in human feces and is transmitted from hand to mouth.

    ◦ Symptoms: No symptoms, live worm in your stool, wheezing, cough, fever, severe abdominal pain, vomiting, restlessness, disturbed sleep.

- Hookworm:

    ◦ How it is transmitted: Hookworm is passed by human feces onto the ground. It is transmitted by walking barefoot on contaminated soil.

    ◦ Symptoms: Diarrhea, barely noticeable abdominal pain, intestinal cramps, colic, nausea, and serious anemia. People in good health may not have any symptoms at all.

- Pinworm infection:

    ◦ How it is transmitted: Found in the colon and rectum, the pinworm infection develops from a pinworms egg. It is transmitted when the female pinworm deposits her eggs in and around the anus. When you touch the eggs with your fingers, the eggs will enter your mouth and travel to your intestines. These eggs are also able to cling to bedding, clothing, toys, doorknobs, furniture, and faucets for up to two weeks. Pinworm is the most common of all the parasitic roundworm infections.

    ◦ Symptoms: No symptoms to very mild symptoms. Itching around the anus or vagina may become intense after the eggs are laid.

- Strongyloidiasis:

    ◦ How it is transmitted: Strongyloidiasis is found in tropical, subtropical, and temperate regions. It is acquired through direct contact of contaminated soil. It enters through human skin, and then makes its way to the intestines.

    ◦ Symptoms: No symptoms to very mild symptoms. Moderate infections may cause burning in the abdomen, nausea, vomiting, and alternating diarrhea and constipation. Severe infections include anemia, weight loss, and chronic diarrhea.

- Trichinosis:

    ◦ How it is transmitted: Unlike other types of roundworms, Trichinosis is not an intestinal infection. It is an infection that affects muscle fibers. It is caused by undercooked sausage, pork, horse, walrus, and bear meat and causes serious problems in muscle fibers. It is transmitted through the consumption of these meats.

- Symptoms: No symptoms to very mild symptoms. Symptoms of the infection in the stomach are diarrhea, abdominal cramps, and tiredness. When the larvae enter the muscle fibers you may feel muscle aches and pains, high fever, swelling in the eyes and face, eye infection and rashes.

- Whipworm:

    - How it is transmitted: Whipworm is contracted by coming in contact with it on your hands, eating food that has come in contact with it, or grown in soil contaminated with it. It is the third most common roundworm to infect humans.

    - Symptoms: There are usually no symptoms. Although, severe infections may cause sporadic stomach pains, bloody stools, diarrhea, and weight loss.

Zoonotic Hookworms

Hookworms are parasitic intestinal nematodes, several of which are zoonotic. In their normal hosts, hookworms may enter the body either by ingestion or through the skin. Larvae that penetrate the skin travel through various organs, including the respiratory tract, before entering the intestines and developing into mature hookworms. Hookworms can cause anemia, abdominal pain and diarrhea when they reside in the intestines or respiratory, dermatologic and other signs during their migration through the body. Young individuals tend to be affected more severely. In cattle, infections may lead to severe disease and pronounced weight loss, with as few as 50 adult worms causing significant anemia in calves. Hookworm disease in cats and dogs can result in anemia, and infections of neonatal pups may prove fatal, even with as few as 50-100 worms present.

Animal hookworm larvae can penetrate the human epidermis, but most species cannot readily enter the dermis, and remain trapped in the skin. These larvae migrate extensively within the skin for a time, resulting in a highly pruritic but self-limited disease called cutaneous larva migrans. One species carried by dogs and cats is increasingly recognized as an intestinal parasite of humans: Ancylostoma ceylanicum has been found in 6-23% of patent human hookworm infections in some parts of Asia. A. caninum also migrates occasionally to the intestines, but usually as a single worm. While one hookworm is unlikely to cause significant blood loss, its presence may result in a painful intestinal disorder called eosinophilic enteritis.

Etiology

Hookworms are nematodes in the superfamily Ancylostomatoidea. In their normal hosts, they are parasites of the intestinal tract. Humans are usually infected by Ancylostoma duodenale and Necator americanus, which are maintained in human populations. Some zoonotic species may also reach the intestines. A. ceylanicum can sometimes be found in large enough numbers to cause typical enteric signs, but A. caninum seems to occur only as a single worm. Rare human intestinal infections with A. malayanum, A. japonica, Necator suillis and N. argentinus have also been reported, but the identification of these organisms is uncertain.

Hookworm larvae that normally mature in the intestinal tracts of animals can cause cutaneous larva migrans in people. Zoonotic hookworms known to cause this condition include A. braziliense, A. caninum, A. ceylanicum, A. tubaeforme, Uncinaria stenocephala and Bunostomum phlebotomum. Other species of hookworms found in animals, including wildlife and captive exotics, might also be able to cause cutaneous larva migrans.

Species Affected

Ancylostoma braziliense is a hookworm of dogs, cats and other carnivores. A. caninum is found in dogs, and has been reported in cats in some parts of Asia and Australia. Uncinaria stenocephala also infects dogs and occasionally cats. A. ceylanicum occurs in wild and domesticated canids and felids, and A. tubaeforme infects cats and other felids. Bunostomum phlebotomum is a hookworm of cattle. Rodents can be paratenic hosts for hookworms including A. braziliense, A. tubaeforme, and U. stenocephala, and possibly other hookworm species.

Ancylostoma duodenale and Necator americanus, species usually found only in humans, have been reported on rare occasions in other mammals.

Zoonotic Potential

A. braziliense is responsible for most cases of cutaneous larva migrans in humans. A. caninum, A. ceylanicum, Uncinaria stenocephala and Bunostomum phlebotomum are involved less frequently, while rare cases have been caused by A. tubaeforme.

A. ceylanicum is the only zoonotic hookworm known to produce patent intestinal infections in humans. A. caninum can cause eosinophilic enteritis, but does not seem to become patent.

Geographic Distribution

A. caninum is the most widespread of all hookworms and can be found in many parts of the world. A. tubaeforme is also widely distributed. A. braziliense is limited to tropical and subtropical regions including Central and South America, the Caribbean and parts of the U.S. A ceylanicum has been reported in parts of Asia, Africa, Australia, and the Middle East and in one publication from Brazil. B. phlebotomum is a parasite of temperate regions, while U. stenocephala occurs in colder climates including Canada, the northern U.S. and Europe.

Transmission and Life Cycle

Adult hookworms live in the intestines. Hookworm eggs shed in the feces are not immediately infective, and hatch in the environment, often within one to a few days. The larvae feed on soil bacteria and molt twice before they become infective third stage larvae. The larvae develop best in warm, moist, sandy soil that is sheltered from direct sunlight. Under optimal conditions, they reach the infective stage in approximately 4 to 7 days. Third-stage larvae that are unable to enter a mammalian host die in approximately 1 to 2 months when their metabolic reserves are exhausted.

Larvae may enter the body either by penetrating the skin, or by ingestion. Penetration of the skin by third stage larvae usually requires at least 5 to 10 minutes contact with contaminated soil. In their natural hosts, these larvae enter the dermis, where they are transported through the lymphatic vessels and veins to the lungs. In the lungs, they penetrate the alveoli and migrate up the respiratory tree to the trachea. They are swallowed and mature into adults in the intestines. Ingested hookworm larvae do not follow this route, but they develop for a period of time in the gastrointestinal wall before re-appearing in the lumen and maturing to adult worms. A. caninum, A. braziliense, A. ceylanicum and B. phlebotomum can either penetrate the skin or be ingested. U. stenocephala is usually acquired by ingestion.

In dogs more than three months old, A. caninum larvae may fail to complete the migration through the lungs and are arrested in the tissues, where they survive as dormant (hypobiotic) larvae. These larvae can move to the uterus or mammary gland during pregnancy, and are transmitted to the pups. This route of transmission does not seem to exist for A. braziliense or U. stenocephala in dogs, or for any hookworms in cats or cattle.

In the intestines, adult hookworms are attached to the mucosa, but change their location every few hours, leaving tiny, bleeding mucosal ulcerations behind. Some species such as A. caninum release a strong anticoagulant that can cause profuse bleeding. Adult hookworms can live for months to a year or more. Reported adult lifespans are approximately 4 to 8 months for A. braziliense, 18 months to 2 years for A. tubaeforme, and about 4 months to a year for U. stenocephala in dogs. The prepatent period varies with the species of parasite, host species and route of exposure. Canine and feline hookworms usually become patent after 13-27 days; however, A. caninum infections transmitted in colostrum or in utero can produce eggs during the second week of life. The prepatent period for B. phlebotomum in cattle is approximately two months. Eggs may be shed intermittently.

Infections in Paratenic Hosts

Paratenic (transport) hosts can be infected orally or through the skin. Larvae do not develop further in paratenic hosts, but become dormant in various tissues. In mice, A. braziliense and A. tubaeforme larvae are mainly found in the head, particularly in the nasopharyngeal epithelium and salivary glands. A. caninum and U. stenocephala occur mainly in the muscles. If a definitive host ingests these larvae, they are released and complete their development to adults.

Infections in Humans

In humans, most zoonotic hookworm larvae cannot penetrate into the dermis. They remain confined to the epidermis, where they migrate for a period of time but eventually die. These organisms cannot be transmitted to others.

A. caninum can occasionally be found in human intestines, while A. ceylanicum seems to occur more frequently. The route of infection with A. caninum is still unknown; either percutaneous or oral transmission may be possible. Infections with this organism do not seem to become patent; only single A. caninum worms have ever been found in humans. A. ceylanicum, however, can mature, mate and produce eggs in people. Patients with intestinal hookworm disease (A. ceylanicum) are not directly contagious to others, as the eggs must develop for a period of time in the soil before they develop into infective third stage larvae. However, the eggs they shed can contaminate the soil. Epidemiological and genetic data suggest that A. ceylanicum may cycle between humans, dogs and cats in some hookworm-endemic communities in Southeast Asia. It is possible that this occurs in other locations as well.

Disinfection

Aqueous iodine at 50-60 parts per million at 15-30 °C (59-86 °F) has been reported to kill hookworm larvae in 15 minutes or less. Other agents which were shown to kill larvae of Necator americanus in a laboratory setting include very hot water (above 80 °C, 176 °F), ethanol (70% for 10 minutes contact time), and Dettol (0.5% for 15 minutes contact time). In the same study, sodium hypochlorite (bleach, 1%) and glutaraldehyde (2%) had no killing effect on N. americanus larvae, and the efficacy of these agents on the larvae of other hookworm species is questionable. Sodium borate (1kg/2 m²) can be used to disinfect the soil. Hookworm larvae are also susceptible to freezing, drying, direct sunlight and temperatures above 45 °C (113 °F).

Incubation Period

The incubation period varies with the number of parasites. Puppies can become symptomatic in the first week of life, before the infection becomes patent.

Clinical Signs

Dogs and Cats

The clinical signs caused by adult hookworms vary with the parasite burden and the age of the animal. They are frequently related to enteritis and/or intestinal blood loss, and are generally more severe in young animals.

In dogs, A. caninum can cause anemia, dark reddish- brown to black hemorrhagic diarrhea, anorexia and dehydration, with associated weakness. Death may occur due to blood loss. The worms can also cause protein and fluid loss and malabsorption, resulting in wasting and decreased growth. Older animals can carry a few worms without clinical signs. Similarly, A. tubaeforme can cause intestinal blood loss, anemia and weight loss in kittens, and large numbers of worms can be fatal. In contrast, U. stenocephala and A. braziliense are not heavy blood-feeders and do not cause anemia or bloody diarrhea. However, they can result in enteric disease, including diarrhea and protein-losing enteropathy.

Larval hookworms may also cause clinical signs during their migration. Dermatitis may be seen where they penetrate the skin. The cutaneous lesions, which can include erythema, pruritus and papules, are usually limited to the feet and often to the interdigital spaces. In some cases, these signs may be severe and result in self-inflicted trauma. Most often, the skin lesions disappear approximately five days after they appear. Large numbers of larvae in puppies can cause pneumonia during their migration through the lungs. Rarely, aberrant larvae in other locations (e.g., the spinal cord) may also become symptomatic.

Cattle

Larval penetration of the lower limbs can cause uneasiness and stamping, and there may be local skin lesions, edema and scabs. The adult worms can cause anemia, rapid weight loss, and alternating diarrhea and constipation. Hypoproteinemia may be seen, but bottle jaw is usually mild. Deaths can occur, especially in calves.

Post Mortem Lesions

Hookworms are small, grayish-white to reddish-white, cylindrical nematodes (approximately 5-20 mm long), found in the intestines. The intestinal mucosa may be congested and swollen, with many tiny hemorrhagic points or ulcers. With many hookworm species, the intestinal contents are bloodstained. In animals with anemia, the liver and other organs may appear pale.

Pneumonia and lung consolidation can be seen with large numbers of larvae in puppies. Skin lesions may be found on the feet, particularly between the toes where the larvae penetrated. Larvae in aberrant sites may be associated with necrotic and hemorrhagic tracts in the tissues, as well as other signs of tissue damage.

Diagnostic Tests

Hookworm infections are diagnosed by fecal flotation and detection of the eggs. Typical Ancylostoma eggs are 55-76 μm in length and approximately 34-50 μm in width, and have a smooth, thin outer shell. They are unembryonated when they are first shed, but develop quickly; at the time of diagnosis, they may contain several cells or a ball of cells. Uncinaria spp. eggs are very similar but slightly larger (70-90 μm x 40-50 μm); they cannot be easily distinguished from Ancylostoma spp. except in mixed infections. First- stage hookworm larvae may appear in preparations from old or stored feces, especially in warm and humid conditions. Eggs are not shed constantly, and repeated sampling may be necessary to detect infections. Coproantigen ELISA tests are in development.

Hookworm larvae can be identified by fecal culture, if identification to the species level is important. Adult worms can be differentiated by their morphology, using published keys. Polymerase chain reaction (PCR) assays have been used in research.

Treatment

Hookworms can be treated with a wide variety of anthelmintics; however, resistance has been detected in the case of some commonly used drugs such as pyrantel in dogs. Supportive care such as supplemental iron, blood transfusions or a high protein diet may also be necessary in some cases.

Prevention

Anthelmintics can be used in ruminants to decrease parasite burdens and pasture contamination. Pasture rotation and other management techniques can also be important components in preventing disease.

Some heartworm preventives may also aid in the prevention of hookworm disease in dogs and cats. Concrete runways, washed at least twice a week in warm weather, should be used for dogs housed in kennels. Clay or sandy runways, as well as soil and lawns, can be decontaminated with sodium borate. To prevent A. caninum infections in puppies, bitches should be free of hookworms, and they should be kept out of contaminated areas during their pregnancy. The dam and puppies should also be housed separately from other animals.

Infections in cats can be decreased by keeping cats indoors and preventing them from eating rodents. Keeping the litter box clean may decrease reinfection.

Morbidity and Mortality

Hookworms are common parasites of cats and dogs. Although their prevalence can vary, species such as A. caninum or A. tubaeforme may infect most of the dogs and cats in some tropical regions. A. ceylanicum is also common in some communities in Asia. In one study, A. ceylanicum was found in 46% of hookworm-infected dogs and cats in a rural community in Malaysia, where nearly 25% of hookworm-positive humans were found to be infected with this same species.

A. caninum and A. tubaeforme infections are generally more serious than U. stenocephala and A. braziliense. Some A. caninum and A. tubaeforme infections may be fatal, particularly in young animals, due to blood loss. B. phlebotomum can cause deaths in calves.

Infections in Humans

Incubation Period

The incubation period for cutaneous larva migrans is short but vaguely established; according to some estimates, it is approximately 1 to 2 weeks. The incubation period for intestinal hookworm disease varies with the number of parasites and can be a few weeks to many months.

Clinical Signs

Cutaneous Larva Migrans

Cutaneous larva migrans is the most common syndrome caused by zoonotic hookworms in humans. Most lesions occur on the legs, buttocks and hands, but they can be found on any part of the body that was exposed to the soil. Initially, there may be a tingling or prickling sensation where the larvae penetrated the skin, followed by a papule at the same location.

Migration of the slow-moving larvae in the skin results in an allergic reaction where they tunnel. The lesions may include papules as well as nonspecific dermatitis, vesicles, or narrow, serpiginous (snakelike), slightly elevated, erythematous lines. The lesions are intensely pruritic, especially at night, and usually advance several millimeters to a few centimeters a day. Pain is occasionally reported, usually in association with vesicles. Secondary bacterial infections can occur due to scratching. Most cases resolve spontaneously in a few days to several weeks, but some untreated lesions have been reported to last for more than a year.

Other lesions are occasionally reported, when larvae penetrate beyond the epidermis. A. caninum larvae may migrate to the muscles, resulting in myositis with persistent swelling and tenderness. These larvae can also cause systemic signs and folliculitis. Ancylostoma spp. larvae have been documented in the eye.

Classic (Intestinal) Hookworm Disease

Although intestinal hookworm disease is usually caused by human hookworms, the zoonotic species A. ceylanicum can also cause this syndrome. With hookworms adapted to humans, the first symptom is usually pruritus at the site of larval penetration. There may also be erythema with small papules or vesicles, which usually persists for 1 to 2 weeks. Migration of the larvae through the lungs may cause coughing and wheezing; however, lung signs are uncommon and are usually mild except with very heavy worm burdens. The adult worms can cause acute intestinal signs such as abdominal pain, nausea, anorexia, vomiting and hemorrhagic diarrhea or melena. Chronic hookworm disease is characterized by blood loss and iron-deficiency anemia, and is associated with fatigue, pallor, tachycardia and dyspnea on exertion. Hypoproteinemia may cause edema, and there can be signs of malabsorption and malnutrition. In children, there may also be adverse effects on physical and intellectual growth. The severity of the disease varies with the worm burden and the amount of blood loss. Heavy infections can be fatal, particularly in infants.

A. ceylanicum infections tend to be milder than those caused by human hookworms. Anemia is usually the most prominent symptom, but other clinical signs, similar to those caused by the human hookworms, are also reported.

Eosinophilic Enteritis

Eosinophilic enteritis is caused by the zoonotic hookworm A. caninum. It is characterized by increasingly severe episodes of abdominal pain associated with peripheral eosinophilia, but no blood loss. Severe cases can mimic appendicitis or intestinal perforation. Some A. caninum infections may be asymptomatic.

Diagnostic Tests

Cutaneous Larva Migrans

A presumptive diagnosis of cutaneous larva migrans is usually made based on the clinical signs. It may be possible to confirm the diagnosis by a biopsy of the affected skin; however, the larvae are rarely found and this test is not usually diagnostic.

Classic (Intestinal) Hookworm Disease

Intestinal hookworm infections, including A. ceylanicum, are diagnosed by identifying the eggs in the feces, as for animals.

Eosinophilic Enteritis

A. caninum infections are difficult to diagnose. The parasite cannot be detected by examination of the feces, as the single worms found in human hosts do not produce eggs. The presence of eosinophilia aids in diagnosis. Ulcerations in the ileum and colon, and occasionally hookworms, may be seen during colonoscopy. Serologic tests for hookworms, including ELISAs and immunoblotting (Western blotting) are used only in research.

Treatment

Intestinal Disease

Hookworms in the intestines can be treated with a variety of anthelmintics. Iron therapy or blood transfusions may also be needed. In hookworm endemic regions, reinfection is likely, in that currently utilized anti-nematode drugs do not confer lasting protection.

Cutaneous Larva Migrans

Cutaneous larva migrans can be treated with topical or oral anthelmintics. Patients with minimal symptoms may not require treatment, as the infection is self-limiting. Pruritus usually decreases within 24 to 48 hours after the initial treatment, and the lesions usually resolve in a week.

Prevention

Zoonotic hookworm disease is best prevented by eliminating the parasites from dogs and cats to decrease contamination of the soil. Dogs and cats should also be kept off beaches and other places where children play in the sand. Sandboxes should be covered when not in use. Sodium borate can be used to sterilize lawns, kennels or other areas. Removing canine feces at least twice a week also decreases soil contamination. The larvae do not survive well in dry, bare soil in direct sunlight.

Prolonged skin contact with contaminated soil should be avoided. Wearing footwear (or gloves when gardening) can decrease the risk of infection. A waterproof sheet may be spread over damp work areas, when working under houses or in other potentially contaminated areas. When visiting the beach, wearing protective footwear, and using waterproof groundcovers rather than towels for sunbathing can help prevent contact with potentially contaminated sand.

Hookworms can be acquired by ingestion as well as skin contact; therefore, unsafe water should be boiled, and potentially contaminated foods avoided. The hands should also be washed before eating and after contact with soil or other sources of hookworms.

Morbidity and Mortality

Cutaneous larva migrans is particularly common in children. The prevalence is also high in adults who have close contact with soil or sand, including gardeners, farmers, miners, exterminators and others whose occupations require crawling under houses. Cutaneous larva migrans is the most common travel-related skin infection in tourists to tropical areas. Infections are self-limiting and the lesions resolve within a few weeks to a few months. Rare cases have lasted up to a year. Deaths are not seen.

Intestinal infections with A. caninum seem to be uncommon. Human intestinal infections with A. ceylanicum are common in some parts of Asia, and possibly other regions. Classic intestinal hookworm disease caused by human hookworms can result in anemia, malnutrition, and, in severe cases, congestive heart failure, severe blood loss or death, especially in young children. The potential for zoonotic intestinal hookworms (A. ceylanicum) to cause severe disease (including non-blood loss related cognitive impairment) or death following natural infection needs further study.

Arthropod Parasites

Arthropods form a huge assemblage of small coelomate animals with "jointed limbs" (hence the name arthro-pods). They exhibit segmentation of their bodies (metamerism) which is often masked in adults because their 10-25 body segments are combined into 2-3 functional groups (called tagmata). They exhibit varying degrees of cephalization whereby neural elements, sensory receptors and feeding structures are concentrated in the head region. Arthropods possess a rigid cuticular exoskeleton consisting mainly of tanned proteins and chitin. The exoskeleton is usually hard, insoluble, virtually indigestible and impregnated with calcium salts or covered with wax. The exoskeleton provides physical and physiological protection and serves as a place for muscle attachment. Skeletal plates are joined by flexible articular membranes and the joints are hinges or pivots made from chondyles and sockets.

Biodiversity

The main arthropod assemblages include crustaceans (crabs, lobsters, crayfish, shrimp), arachnids (spiders, scorpions, ticks, mites) and insects (beetles, bugs, earwigs, ants, bees, termites, butterflies, moths, crickets, roaches, fleas, flies, mosquitoes, lice). Most parasitic arthropods belong to 2 main classes: the 6-legged insects, and the 8-legged arachnids.

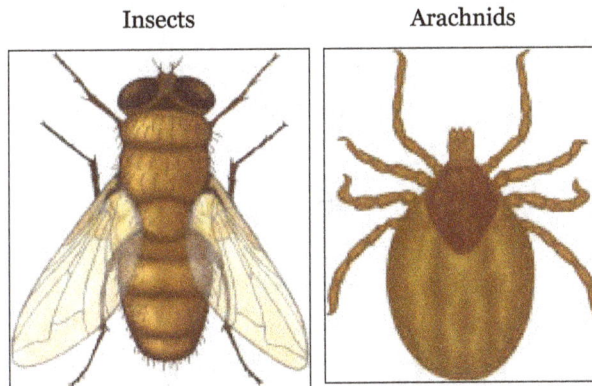

Insects   Arachnids

- Insects have 3 distinct body parts, commonly called the head, thorax and abdomen. The head has 2 antennae and the thorax has 6 legs arranged in 3 bilateral pairs. Many insect species also have 2 pairs of wings attached to the thorax. Parasitic insect species include fleas, flies and lice which actively feed on host tissues and fluids at some stage in their life-cycles.

- Arachnids have 2 body parts known as the prosoma (or cephalothorax) and opisthosoma (or abdomen). The cephalothorax has 8 legs arranged in 4 bilateral pairs and arachnids do not have wings or antennae. Important parasitic assemblages include the ticks and mites which bite into tissues and feed off host fluids.

Collectively, arthropods account for a substantial share of global biodiversity, both in terms of species richness and relative abundance. There are over 1,000,000 species of insects and over 50,000 species of arachnids. They are very successful and adaptable organisms and are capable of forming large populations due to their rapid and fertile reproduction rates. Many species are also able to withstand adverse environmental conditions by undergoing periods of developmental arrest (diapause). The protection afforded by their exoskeletons allows them to colonize many habitats and they overcome the problem of growing larger in a non-expandable exoskeleton by undergoing periodic moulting (or ecdysis) which is mediated by hormones. Developmental stages between moults are referred to as instars. Moulting is a complex process and its timing is mediated by many environmental and physiological cues. It involves detachment of the hypodermis from the procuticle, partial resorption of the old cuticle, production of a new epicuticle, dehiscence (splitting) of the old cuticle, emergence of the animal, stretching and expansion of the new cuticle by air and/or water intake, and then sclerotization of the new cuticle.

Life-cycles

Adult arthropods are generally small in size, most are visible but some remain microscopic. Arthropod sexes are separate and fertilization is internal. A wide range of mating behaviours, insemination and egg production strategies are involved. In most species, the egg develops into a larva: i.e. a life-cycle stage that is structurally distinct from the adult and must undergo metamorphosis (structural reorganization) before becoming an adult. This metamorphosis may be complete (involving major changes during a pupation stage) or incomplete (involving gradual changes in nymph stages). For example, the grub-like larval stages of flies and fleas form cocoon-like pupae where they undergo complete metamorphosis and emerge as radically-different adult insects. In contrast, the larval instars (or nymphs) of lice, ticks and mites undergo incomplete metamorphosis through a series of moults gradually becoming more adult-like in appearance.

Complete metamorphosis   Incomplete metamorphosis

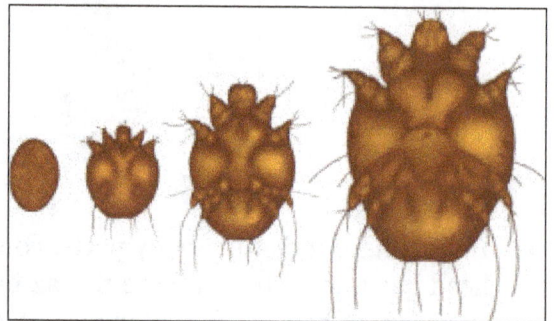

Arthropods are involved in nearly every kind of parasitic relationship, either as parasites themselves or as hosts/vectors for other micro-organisms (including viruses, bacteria, protozoa and helminths). They are generally ectoparasitic on, or in, the skin of vertebrate hosts. Many species are haematophagous (suck blood) while others are histophagous (tissue-feeders) and bite or burrow in dermal tissues causing trauma, inflammation and hypersensitivity reactions. Infestations are transmitted from host-to-host either by direct contact or by free-living larvae or adults actively seeking hosts.

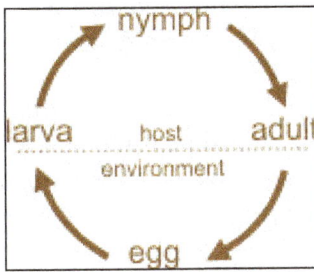

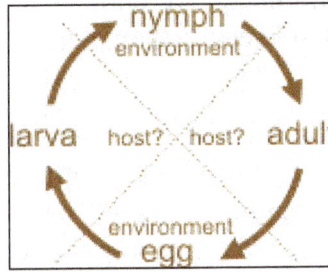

| Direct | host-seeking (all feeding stages parasitic) | host-seeking (larva or adult parasitic) |

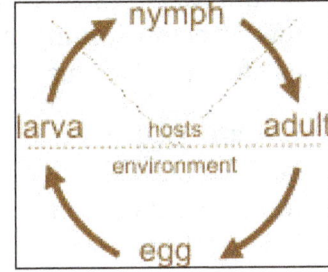

- Direct transmission of infective stages occurs when hosts come into close contact with each other or share quarters, bedding or clothing. Larvae, nymphs or adults may cross from one host to another, while eggs or pupae may contaminate shared environments. Insects (fleas and lice) and arachnids (mites) rely on close contact between hosts.

- Many adult insects actively seek hosts in order to feed or lay eggs. Winged insects (mosquitoes, flies) fly to new hosts to feed while fleas jump onto passing hosts. Some adult flies (botflies) do not feed on their hosts but deposit eggs from which larvae emerge and feed on host tissues and exudates.

- Tick larvae actively seek hosts by climbing vegetation and questing for passing hosts. Some species complete their life-cycle on the same host (one-host ticks) while others detach after feeding and drop to the ground to moult before seeking new hosts as nymphs or adults (two-host or three-host ticks).

Taxonomy

Insects exhibit extraordinary biodiversity, both in terms of species richness (numbers of species) and relative abundance (population sizes). Most parasitic species belong to three main groups: the jumping fleas (Siphonaptera); the winged flies (Diptera); and the wingless lice (Phthiraptera).

- Fleas are bilaterally-flattened wingless insects with enlarged hindlimbs specially adapted for jumping (up to 100 times their body length). Jumping feats are accomplished using elastic resilin pads which expand explosively when uncocked from the compressed state. Fleas undergo complete metamorphosis whereby grub-like larvae form pupae from which adult fleas emerge. The larvae are not parasitic but feed on debris associated mainly with bedding, den or nest material, whereas the adult stages are parasitic and feed on host blood. There are some 2,500 flea species, most parasitic on mammals (especially rodents) and some on birds. They vary in the time spent on their hosts ranging from transient feeders (rodent fleas) to permanent attachment (sticktight fleas and burrowing chigoes).

- Flies and mosquitoes are winged insects with two pairs of wings attached to the thorax and a well-developed head with sensory and feeding organs. They undergo complete metamorphosis involving a pupation stage. Different species vary in their feeding habits, both as adults (parasitic or free-living) and larvae (parasitic or free-living). There are over 120,000 species belonging to 140 families. Two main suborders are recognized on the basis of structural differences, Nematocera (adult stages parasitic, larval stages often free-swimming) and Brachycera (adult stages parasitic or free-living, larvae stages often predaceous).

- Lice are small wingless insects, dorsoventrally flattened, with reduced or no eyes and enlarged tarsal claws for clinging. All lice undergo gradual metamorphosis and there are no free-living stages. Eggs are cemented to hair/feathers whereas nymphs and adults cling to hair/feathers. Two orders of lice are recognized on the basis of their mouthparts: the Mallophaga (chewing/biting lice) with some 3,000 species infesting birds and mammals; and the Anoplura (sucking lice) with 500 species found on mammals.

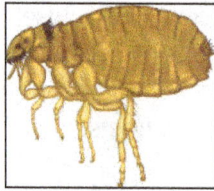      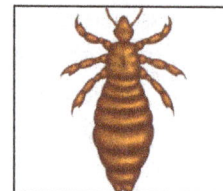

       Fleas   Flies   Lice

Many non-spider arachnids (subclass Acari) are found as parasites on animal or plant hosts. They belong to two main groups: the macroscopic ticks and the microscopic mites. Many species are important in human and animal medicine as causes of disease or as transmission vectors for other pathogens.

- Ticks are epidermal parasites of terrestrial vertebrates that may cause anaemia, dermatosis, paralysis, otoacariasis and other infections (transmit viral, bacterial, rickettsial, spirochaete, protozoal and helminth pathogens). They feed mainly on blood and their mouthparts are armed with small backward-facing teeth to aid in attachment. All ticks undergo gradual/incomplete metamorphosis whereby larval and nymphal instars resemble adults. The integument is relatively thick and respiration occurs via spiracles (usually only one pair) and trachea. Two major families of ticks are recognized on the basis of many morphological features: the Ixodidae (hard ticks with a tough cuticle and a large anterodorsal scutum) with some 650 species that infest mammals, birds and reptiles; and the Argasidae (soft ticks with a leathery integument and no scutum) with 160 species that infest mainly birds and some mammals.

- Mites are microscopic arachnids which undergo gradual or incomplete metamorphosis. Adults and nymphs have 4 pairs of legs whereas larvae have 3 pairs. Over 30,000 species of mites have been described, many are free-living species, some are plant parasites while others are parasitic on terrestrial and aquatic hosts. Most parasitic species feed on skin debris or suck lymph, some burrow into the skin, some live in hair follicles, and some in the ear canals. Their mouthparts are variable in form but the hypostome is never armed with teeth. The integument is usually thin and three orders are recognized on the basis of their respiratory systems: the Mesostigmata with respiratory spiracles (stigmata) near the third coxae; the Prostigmata (Trombidiformes) with spiracles between the chelicerae or on the dorsal hysterosoma; and the Astigmata (Sarcoptiformes) without tracheal systems as they respire through the tegument.

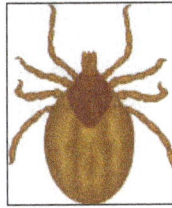

Ticks

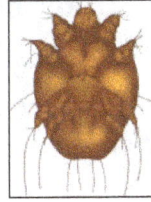

Mites

Acanthocephala

The Acanthocephala are endoparasitic worms of slender cylindroid or slightly flattened form and hollow construction. They live as adults in the intestine of vertebrates and as larvae in arthropods. The diagnostic feature of the phylum is the organ of attachment consisting of an invaginable proboscis that forms the anterior end.

This proboscis is armed with rows of recurved hooks. The body wall consists of cuticle, syncytial epidermis permeated with spaces and sub-epidermal musculature. In connection with the proboscis apparatus the epidermis forms two elongated bodies termed lemnisci that hang down into the trunk.

Mouth, anus and digestive tube are completely wanting. There is no circulatory system. Excretory organs when present are of the nature of protonephridia and open into the terminal part of the reproductive system.

The nervous system consists of a ganglion near the proboscis and two lateral cords extend posteriorly from the ganglion along with numerous minor nerves. The sexes are separate, the females are generally larger than the males, the males are provided with a copulatory apparatus and the terminal part of the female apparatus is also somewhat complicated.

The eggs develop within the maternal body into a larva that requires an intermediate invertebrate host for its further development. There are over 500 known species. Echinorhynchus is the chief genus of Acanthocephala.

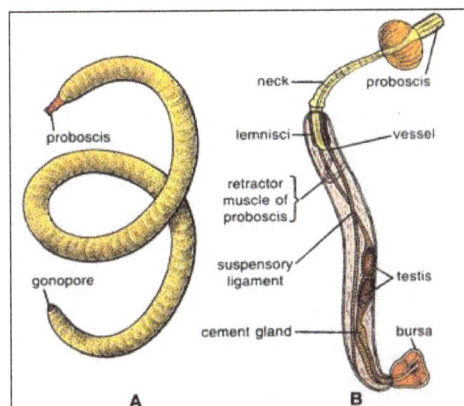

Echinorhynchus gigas, A-Male; B-Female

Habit, Habitat and External Characters of Acanthocephala

The genus Echinorhynchus is a common parasite in the intestine of mammals, birds, reptiles, amphibians and fishes. The largest species E. (Gigantorhynchus) gigas is found in the pig and has once been recorded in the human being. It may attain the length of 50 to 65 cm.

External Characters of Acanthocephala

The body is cylindrical and ends in front in a protrusible portion, the proboscis, which is cylindrical and is covered with many rows of recurved chitinous hooks.

In many species the body is ringed or constricted at regular intervals and so presents a more or less segmented appearance. Such segmentation reaches the highest degree in certain species of Moniliformis and Mediorhynchus.

In all these forms it is difficult to distinguish dorsal from ventral surface from external appearance when the body is curved, the concave surface is ventral and when the proboscis is covered with unequal size of hooks, the larger hooks are ventrally located, but in most cases dorsoventrality can be the proboscis includes the receptacle or sheath into which it invaginates or withdraws the lemnisci and the main ganglion of the nervous system. The neck is generally short but sometimes much elongated.

The trunk may be cylindroid, flattened, curved or coiled with a smooth, wrinkled or segmented surface. In several genera the trunk is more or less noticeably differentiated into a broader fore trunk and a slender hind trunk in varying proportions in different species. There is no trace of mouth, anus or excretory pore. The gonopore occurs at or near the posterior extremity.

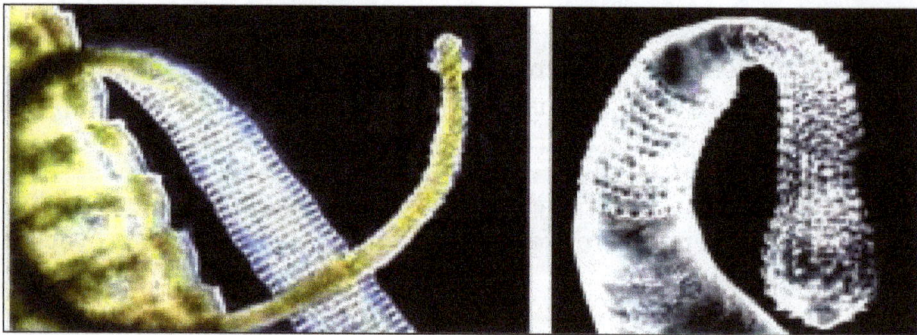

Moniliformis: An Acanthocephalan

Body Wall of Phylum Acanthocephala

The body wall is covered with stout cuticle of homogeneous structure. Beneath cuticle lies the remarkable epidermis or hypodermis. The epidermis is a thick layer of fibrous syncytial construction comprising three fibrous strata an outer layer, only slightly thicker than the cuticle, of parallel radial fibres, a middle somewhat thicker felt work of layers of fibres running in different directions and an inner layer of radial fibres.

The inner layer is the thickest of the three and by many authors is regarded as the epidermis proper, while the outer radial and the felt layers are assigned to the cuticle.

The fibres of the epidermis do not appear to be the nature of connective tissue but rather seem to be protoplasmic strands. There are no indications of cell walls and the entire epidermis forms a syncytium. The nuclei and the lacunar systems are situated in the inner radial layer.

In Acanthocephala the number of nuclei is approximately constant for each species, at least in early stages, and in many families throughout the life.

These nuclei range in shape from globose or oval to rosette, amoeboid or highly branched forms and are of relatively large size up to 2 mm or more in length. The inner radial layer of epidermis contains the lacunar system, a set of channels without definite walls but having a more or less definite pattern.

The lacunar system consists of two longitudinal vessels with regularly spaced transverse connections. The main channels in this case are medially located, i.e., they are dorsal and ventral. The lacunar system is confined mainly to the epidermis and it does not communicate with the exterior or with any other body structure.

It contains a nutritive fluid presumably obtained by absorption Dissection of male, from the host and, therefore, serves as a food-distributing system in the absence of any other. The nutritive fluid in the system moves only with body movements.

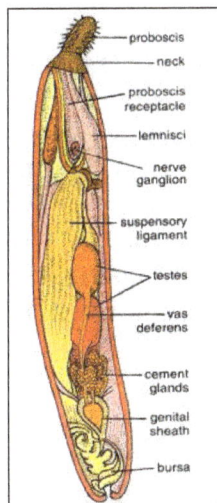

Echinorhynchus: Dissection of Male.

The epidermis is followed by a thin layer, the dermis (also called basement or binding layer). It also permeates the underlying musculature. The muscle layer forming the innermost coat of the body wall is relatively thin consisting of outer circular and inner longitudinal fibres.

These muscle layers are also syncytial forming a fibrous network and probably contain a more or less definite number of nuclei. The muscle fibres consist of a cytoplasmic and a fibrillar portion; the fibres may run along one side of the cytoplasmic part or may encircle the latter. No definite lining membrane bounds the body wall from the pseudocoel.

Proboscis Apparatus of Acanthocephala

The everted proboscis varies from a short cylindrical to globose shape to a long cylinder. It is armed with recurved hooks that are similar in shape but varied in size and arrangement. They are most commonly arranged in alternating radial rows but may be concentric or irregular; they may be of even size over the proboscis.

The hooks may be largest at the summit and decrease gradually toward the base or often are largest in the middle region decreasing towards the end. The number of rows, the number per row, and the size pattern of the hooks are fairly constant for each species, and in fact the proboscis armature constitutes one of the most important taxonomic characters of the phylum.

The larger hooks have roots sunk into the proboscis wall and some authors reserve the term hooks for these, calling the smaller, rootless members of the armature spines. The hooks and spines are of unknown chemical nature but apparently consist of the same material as the dermis from which they seem to originate. They are covered with cuticle.

The proboscis is invaginable and withdraw-able into a muscular sac, the proboscis receptacle that is fastened in a circle to the inner surface of the proboscis wall and hangs into the pseudocoel.

It is composed in some groups of a single muscle layer, in others of two layers. Proboscis and receptacle are operated by special muscle bands. From the proboscis summit there extends posteriorly along the proboscis interior a muscle or group of muscles termed the retractor, invaginator or invertor muscle.

This inserts on the receptacle wall and also passes through this wall to continue as the dorsal and ventral receptacle retractors that terminate on the trunk wall. In the more complicated types there are also a number of dorsal, ventral and lateral receptacle protrusors that originate in a circle from the neck wall and insert on the rear part of the receptacle.

Finally generally distributed in the Acanthocephala there are the retractors of the neck that originate near the posterior boundary of the neck and insert on the trunk. The retractors of the neck encircle all the other proboscis muscles and embrace the lemnisci as compressors of lemnisci.

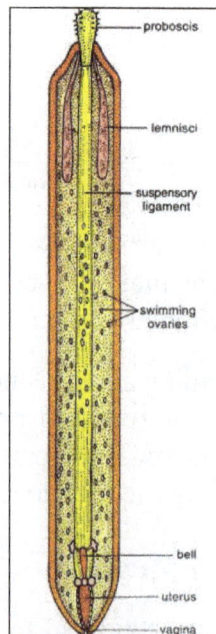

Echinorhynchus: Dissection of Female.

From the posterior part of the neck region there extend posteriorly into the trunk pseudocoel a pair of projections of the inner radial layer of the epidermis known as lemnisci. The lemnisci are generally long slender bodies; they are supplied by vessels of the lacunar system and contain either a limited and definite number of fragmented nuclei.

They are externally covered by the dermis and are well supplied with the lacunar system. At certain levels the lemnisci are enclosed in the neck retractors. The function of the lemnisci is that they act as reservoirs for the fluid of the lacunar system of the presoma when the proboscis is invaginated.

Nervous System of Acanthocephala

The nervous system consists of the cerebral ganglion, the branches from this and in the male a pair of genital ganglia with branches. The cerebral ganglion, often called simply the ganglion, is a large cellular mass enclosed in the proboscis receptacle in contact with its ventral walls. It consists of a central fibrous mass embraced by ganglion cells, 86 in Macracanthorhynchus, 80 in Hamanniella and 73 in Bolbosoma.

In the first two genera, the ganglion gives off two single and three pairs of nerves, in Bolbosoma one single and five pairs of nerves.

The former comprise an anterior median and a ventral anterior nerve to the musculature and sensory papillae of the proboscis, a pair of lateral anterior nerves to the lateral protrusors, a pair of lateral medial nerves to the receptacle wall, and the pair of main lateral posterior nerves that proceed to the posterior end of the animal.

The lateral posterior nerves pierce the receptacle wall, proceed laterally to the body wall of the trunk into which they give branches, and then run posteriorly in the lateral body wall in the longitudinal muscle layer to the posterior end, giving off genital branches in females.

In males, however, branches from these nerves enter a pair of genital ganglia situated in the penis base and connected with each other by a ring commissure. From the genital ganglia branches proceed anteriorly along the male genital tract and posteriorly into the bursa where some terminate in bulbous sense organs.

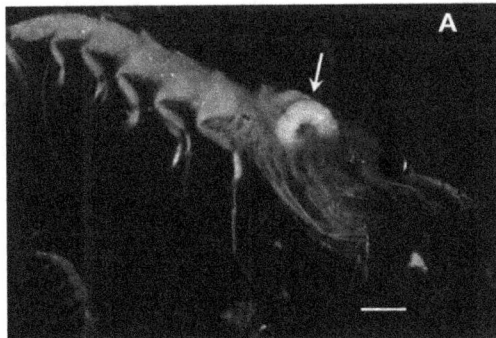

Bolbosoma

Sense Organs of Acanthocephala

Organs of special sense are not developed in correlation to entoparasitic life.

The known sense organs comprise three in the proboscis and several in the male bursa and penis. In the proboscis there is a sensory organ in the centre of the tip and in some genera one on each side in the neck. The terminal sense organ consists of a small pit beneath which there is a fusiform nerve ending of a nerve fibre that makes a coil just below its termination.

The pair of lateral proboscis sense organ is similar except that several coiled nerve fibres are involved.

These proboscis sense organs are supplied by certain of the anterior nerves coming from the cerebral ganglion. In males fibres from the genital ganglia terminate in bulbous or spherical sense organs of which there are seven or eight around the rim of the bursa and a number in the penis. It is usually supposed that all the acanthocephalan sense organs are of tactile nature.

Excretory Organs of Acanthocephala

The excretory organs consist of a pair of small bodies, protonephridia, situated at the posterior end near the genital aperture. In most genera each protonephridium consists of a branching mass of flame bulbs attached to a common stem.

The number of flame bulbs in each protonephridium ranges from about 250 to 700. The flame bulbs are devoid of nuclei and are, therefore, not cells; usually three nuclei occur in the main branches or in the wall of the chamber.

The flame bulb consists of a linear row of cilia. In certain genera (Oligacanthorhynchus, Nephridiorhynchus), the flame bulbs open directly into a sac from which the nephridial canal leads.

In any case the two canals unite to a single canal or to a bladder and this joins the common sperm duct in the male and uterus in the female. The terminal canals of the reproductive system are then urinogenital canals in the Archiacanthocephala.

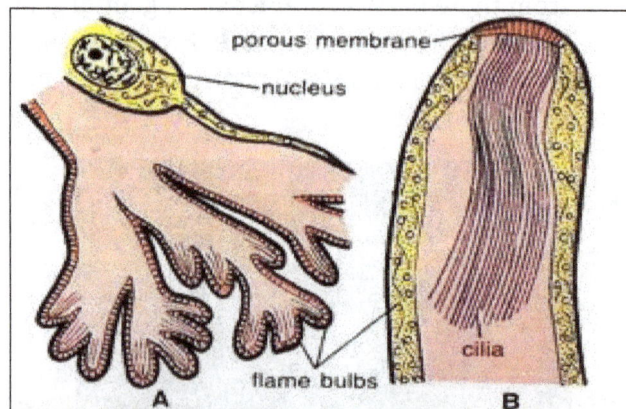

Echinorhynchus: A-L.S. through terminal twigs of a protonephridium; B-L.S. of a flame cell.

Ligament Sacs and Ligament Strand of Acanthocephala

These are structures peculiar to the Acanthocephala. The ligament sacs (formerly called ligaments) are hollow tubes of connective tissue with or without accompanying muscle fibres that run the length of the body interior and enclose the reproductive organs.

Anteriorly they are attached to the posterior end of the proboscis receptacle or the adjacent body wall and posteriorly they terminate on or in some part of the reproductive system. In the female there are two ligament sacs, a dorsal and a ventral, whose medial walls make contact in the frontal plane and which communicate anteriorly by an opening.

The dorsal ligament sac attaches posteriorly to the anterior rim of the uterine bell, the ventral sac extends to the posterior end of the body, opening en route into the ventral aperture of bell. In males there is only one sac, the ventral sac is wanting. The dorsal sac encloses the testes and the cement glands and posteriorly becomes continuous with the genital sheath.

Haffner has reported the ligament strand, a nucleated strand found between the two ligament sacs when present or situated along the ventral face of the single ligament sac. The gonads in both sexes are attached to this strand. According to Haffner's analysis, the ligament strand represents the endoderm mid-gut. The ligament sacs are regarded by Hyman as separated parts of the pseudocoel.

Pseudocoel of Acanthocephala

The pseudocoel is a cavity, not provided with any lining membrane, between the body wall and the ligaments. It is small in forms with two ligament sacs but attains considerable size in those with one sac only. It also extends into the presoma between the muscle bands.

As the cavity lacks a lining membrane, it is obviously not a coelom. As the ligament strand apparently represents the endoderm, the body cavity is a space between the endoderm and the body wall and, hence, classified as a pseudocoel. The pseudocoel is filled with a clear fluid.

Reproductive System of Acanthocephala

The greater part of the body is occupied by the reproductive organs. The sexes are separate and the female is larger than the male. In both the sexes the gonads and their ducts are connected with a ligament strand which extends backwards from the end of the proboscis sheath.

In males there are two oval, rounded or elongated testes enclosed in the ligament sac and attached to the ligament strand. From each testis a sperm duct proceeds posteriorly inside the ligament sac. Small enlargements representing spermiducal vesicles may occur along the sperm ducts.

A cluster of unicellular gland cells known as cement glands (usually six or eight in number and of variable shape-rounded, pyriform, clavate or tubular) open into the sperm duct shortly behind the more posterior testis. The ducts of these cement glands, either separately or after union into one or two main ducts, enter the common sperm duct.

The sperm ducts, the cement ducts and the protonephridial canals (when present) are all enclosed in a muscular tube, the genital sheath.

The genital sheath terminates on the muscle cap of bursa. Inside the genital sheath, the two sperm ducts unite to a common sperm duct which may present a saccular enlargement, the seminal vesicle, the cement duct enter the common sperm duct and the common protonephridial canal, when present, also unites with common sperm duct.

The urogenital canal so formed penetrates the centre of the penis, a short conical protrusion.

The penis projects into hemispherical or elongated cavity the bursa that is eversible to the exterior and grasps the rear end of the female in copulation. The bursa is composed of internal body wall of which the muscular layer is greatly thickened in the proximal part of the bursa, forming the muscular cap. The sperms are long filaments without definite heads.

The female reproductive system departs from the usual in many ways. The original single or double ovary breaks up into fragments termed ovarian walls that float free in the dorsal ligament sac but as the latter sac soon ruptures the balls occupy the pseudocoel. The ligament sacs lead to the first part of the female canal termed the uterine bell.

The uterine bell is a muscular, funnel- shaped or tubular organ that by peristaltic contractions engulfs the developing eggs and passes them onward.

The bell has a single (or a pair of) posterior ventral openings through which the immature eggs, which are spherical, pass back into the body cavity. At its posterior end the bell narrows to a uterine tube composed of several large cells with conspicuous nuclei and bearing two bell pouches that extend anteriorly.

The uterine tube enters the uterus a muscular tube of some length and this is followed by the short non-muscular vagina opening to the exterior.

The nephridia lie alongside the uterine bell, the two protonephridial ducts run in the dorsal wall of the bell and the common canal formed by their union opens into the beginning of the uterine tube. The ovarian balls consist of central syncytium from which ovogonia separate, passing to the periphery for further development.

Copulation and Development of Acanthocephala

In copulation the everted male bursa grasps the posterior end of the female and the penis enters the vagina and discharges sperms into the uterus. This is followed by the discharge of the cement secretion which sets as a plug in the gonopore and as a cap over the whole posterior tip, preventing escape of the sperms.

The mature ova are elliptical and surrounded by a membrane. Fertilisation takes place in the body cavity and after fertilisation a membrane arises inside the original egg membrane.

In the meantime the eggs have escaped from the ovarian balls and continue development in the pseudocoel or inside the dorsal ligament sac until a larval stage provided with a rostellum armed with hooks is reached. Meanwhile, a third membrane, usually termed shell, has formed between the two membranes already present around the embryo.

These ovic larvae are engulfed by the uterine bell and passed towards the uterine tube; those not sufficiently mature may be returned through the ventral bell aperture into pseudocoel or ventral ligament sac. The ripe ones proceed into the bell pockets and then along the uterus and vagina to the exterior. These elliptical ovic larvae must be ingested by the proper invertebrate host before they can develop further.

Parasitic Mites

People who live or work in structures that are rat infested or that house nesting birds are frequently attacked by parasitic mites that migrate from their nests into the occupied portion of the building. Bites from these parasites can cause painful dermatitis and severe itching. Bird and rodent mites normally live on the host or in their nests, but will migrate to human occupied areas of the structure if their host dies or leaves the nest. Oftentimes exterminators will eliminate an infestation of mice or rats causing a severe nuisance of mites for the structure's humans or animal pets. These pests are very tiny, but can generally be seen with the naked eye. They are the size of a small dot, similar to a period at the end of a sentence.

Rodent Mites

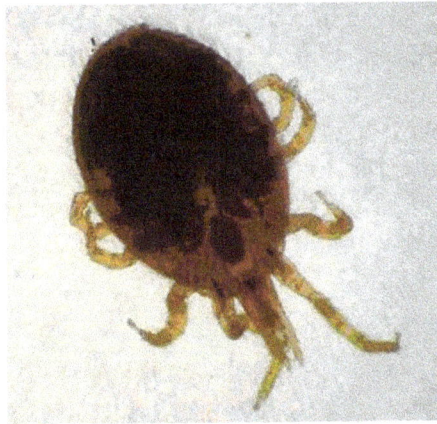

There are several types of rodent mites that typically feed on humans. The house mouse mite (*Liponyssoides saguineus*) prefers mice but will also bite rats and humans, often causing a rash around the bite. The tropical rat mite (*Ornithonyssus bacoti*) are extremely active and will crawl long distances to find a blood meal. They most often make their presence known shortly after control measures have occurred to eliminate a rat infestation and quickly migrate to occupied areas of the structure to find an alternative human or animal host. The tropical rat mite's bite is painful and causes skin irritation and itching. The spiny rat mite (*Laelaps echidnina*) will hide in areas near nests and resting areas of rats and feed by night.

Fowl Mites

These mites are similar in size, appearance and behavior to tropical rat mites. They are generally most troublesome in summer during and shortly following the nesting season. Bird mite problems occur when sparrows, pigeons, or mourning doves build their nests in occupied structures. During the period the birds and their young occupy the nest the mite population is healthy and increases in size. When the birds fledge and leave the nest the mites migrate into the structure seeking alternative hosts. These types of migrations can cause an excessive number of bites experienced by occupants of the infested structure. Intermittent bites may also be experienced by building occupants in structures in which birds, such as pigeons, may congregate on roofs or other portions of the structure as resting sites.

Chicken Mites

(Dermanyssus gallinae) primarily infests chickens, but may also feed on sparrows, pigeons and blackbirds. Chicken mites can survive for significant amount of time away from its host. It hides in cracks and crevices during the daytime and feeds at night. The northern fowl mite (*Ornithonyssus sylviarum*) can survive for more than a month away from its host. Cheyletiella ("walking dandruff") mites may also infest birds, but are more frequently found on pets such as dogs, cats and rabbits. They can cause a mange-like condition in animals and itching in people who handle the infested pets. The Cheyletiella mite does not live off the host for very long.

Human Scabies or Itch Mites

The Sarcoptes scabiei are mites that infest mammals, including humans. The human scabies mites are very small and are rarely seen. When they first burrow into the skin they cause very little irritation, but within a month a rash will begin to appear and severe itching is experienced. Because of this delay in symptoms of a month after infestation it is possible other members of the household or institution are also infected. Human scabies are very contagious and if an infestation occurs in a facility such as a nursing home it is likely to rapidly spread. Although human scabies will not live beyond a few days outside of a host, it is logical that coincident with medical treatment, a Therma-Pure treatment of the affected area be performed. This treatment would disinfect porous materials such as chairs, lounges and beds that might harbor the mite.

References

- Helminth-introduction: parasite.org.au, Retrieved 2 April, 2019

- 58959-types-parasitic-nematodes: livestrong.com, Retrieved 10 July, 2019

- Tapeworms, pathology, diseases-and-conditions, medicine: encyclopedia.com, Retrieved 7 August, 2019

- Flatworm, animal: britannica.com, Retrieved 5 June, 2019

- 15240-roundworms, diseases, health: clevelandclinic.org, Retrieved 3 February, 2019

- Hookworms: iastate.edu, Retrieved 10 July, 2019

- Arthropod-intoduction: parasite.org.au, Retrieved 20 January, 2019

- Acanthocephala-characters-sense-organs-and-affinities, invertebrate-zoology: biologydiscussion.com, Retrieved 11 May, 2019

- Parasitic-mites, mites, pest-services: thermapure.com, Retrieved 14 June, 2019

Chapter 5

Life-cycles and Host-Parasite Interactions

The life cycle of parasites involves numerous stages where it looks for a host and then infects it in order to derive nutrition. The study of host-parasite interactions plays an integral role in exploring the biological processes governing parasitic diseases. The topics elaborated in this chapter will help in gaining a better perspective about the life cycles of parasites and the host-parasite interactions.

Essential aspects to a Parasite Life Cycle

Find a Host

1. Active: Host produces either chemical, thermal or light signals to which the infective stage is sensitive.

2. Passive: Infective stages are dispersed passively through environment, such as waves or water currents.

Enter a Host

1. Active: Infective stage may burrow into skin, as cercariae of blood flukes do.

2. Passive: Host will ingest or inhale infective forms.

Overcome Host Defenses

Mechanisms include:

- Antigen Shielding: Surface of parasite adsorbs host derived antigen, so that parasite is recognized as "self". (Documented among adult blood flukes).

- Surface Antigen Shifting: Proteins forming protein surface change so that immune reactions lag behind development of the parasites. (Documented in Trypanosoma gambiense and T. rhodesiense (African trypanosomiasis).

Derive Nutrients from Host

1. Aerobic Metabolism

2. Obligate Anaerobic Metabolism.

Reproduce more Individuals

1. Hermaphroditism (especially simultaneous),

2. Polyembrony,

3. Very high egg output: among some tapeworms, daily output can be in millions, but parental care is negligible.

Disperse Young to New Hosts

1. Presence of Obligate Free-living Stage: Needs to tolerate changes in temperature, osmotic pressure, desiccation.

2. Passive or Active Dispersal: Cercariae of digenetic trematodes burrow out of snail host in search of 2nd intermediate host or of definitive host.

3. Parasite-Induced Change in Host Behavior: Intermediate stages will induce qualitative changes in the intermediate host such that the infected intermediate host is more likely to be captured by a predator which is the definitive host.

4. Parasite-induced change in host morphology, making them less able to avoid predators, as in the case of Ribeiroia ondotrae metacercariae among green frogs.

Effects of Parasitic Life on Parasites and Host

In early parasitism there are no morphological changes in the parasite though physiological adaptations take place, later the following changes occur:

(a) There is a reduction in the organelles of locomotion's, since the parasites are transported by the host, so that the locomotor organelles are simplified and finally lost. In some intestinal Sporozoa (Gregarina) only metaboly occurs, but in intracellular parasites (Plasmodium) there is no locomotion.

(b) The form and shape of the body become very simple with no complex organelles (Plasmodium).

(c) Organelles of fixation appear in some intestinal parasites (Gregarina).

(d) Organelles of nutrition are simplified (Balantidium) or even lost (Plasmodium) since food is absorbed by the body surface.

(e) Parasites acquire ability for rapid multiplication to form numerous young ones, this ensures that at least some of the offspring will find a suitable host and continue the species (Plasmodium).

(f) Many have two hosts in their life cycle, and one of the hosts also acts as a vector to disseminate the parasite (Trypanosoma in man and tsetse fly).

Host Specificity

Some parasites are restricted to a few hosts only, e.g., Gregarina in a few insects, or Opalina in Anura only, but some parasites have become adapted to a large variety of hosts, e.g., Trypanosoma is found in all classes of Vertebrates in which it parasitises some five hundred species.

Thus, in the development of host relationship the above two general trends are seen, this is due partly to the infective powers of the parasite, and partly to the degree of susceptibility of the host.

Effects of Parasitism on the Host

The following pathological conditions may be caused by parasites in their hosts:

- Destruction of cells and tissues of the host may take place by movement or feeding activities of the parasite, e.g., Entamoeba histolytica eats the tissue cells of the colon and red blood corpuscles of the host; Plasmodium feeds on liver cells and erythrocytes.

- Parasites may cause enlargement and disorders of lymph glands, spleen and liver, e.g., Leishmania or parasites may cause ulcers in the intestine, liver and brain, e.g., Entamoeba.

- Parasites may secrete poisonous toxins which cause some disease in the host, e.g., Plasmodium causes malaria.

But in most cases of parasitism there is a mutual adaptation between the host and the parasite, the parasite is able to live and reproduce without any apparent injury, and the host offers a resistance or acquires an immunity against the parasite by producing antibodies which neutralize the effects of the parasite, or by becoming immune due to previous infection, or by increasing its powers of repairing and regenerating the injured tissue cells.

At times the host destroys the parasite by phagocytosis with the aid of leucocytes or cells of the spleen, bone marrow and liver.

The host may succeed in destroying the parasite or it may remain infected but become immune, so that it becomes a carrier of the parasite. Generally there is a delicate adjustment between the parasite and the host and they come to an elaborate compromise, if this mutual adjustment is lacking, then either the parasite is killed or the host is destroyed.

Life Cycle of the Malaria Parasite

The malaria parasite has a complex multistage life cycle occurring within two living beings, the vector mosquitoes and the vertebrate hosts. The survival and development of the parasite within the invertebrate and vertebrate hosts, in intracellular and extracellular environments, is made possible by a toolkit of more than 5,000 genes and their specialized proteins that help the parasite to invade and grow within multiple cell types and to evade host immune responses. The parasite passes through several stages of development such as the sporozoites (Gr. *Sporos* = seeds; the infectious form injected by the mosquito), merozoites (Gr. *Meros* = piece; the stage invading the erythrocytes), trophozoites (Gr. *Trophes* = nourishment; the form multiplying in erythrocytes), and gametocytes (sexual stages) and all these stages have their own unique shapes and structures and protein complements. The surface proteins and metabolic pathways keep changing during these different stages that help the parasite to evade the immune clearance, while also creating problems for the development of drugs and vaccines.

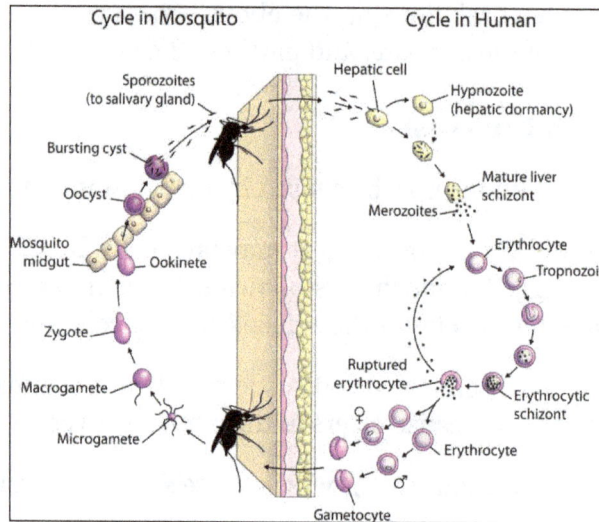

Life Cycle of Malaria Parasite

Sporogony within the Mosquitoes

Mosquitoes are the definitive hosts for the malaria parasites, wherein the sexual phase of the parasite's life cycle occurs. The sexual phase is called sporogony and results in the development of innumerable infecting forms of the parasite within the mosquito that induce disease in the human host following their injection with the mosquito bite.

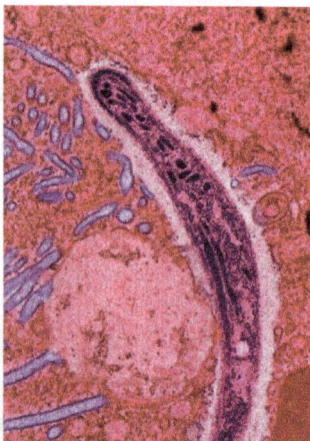

Coloured TEM of malaria sporozoites in a Anopheles mosquito gut.

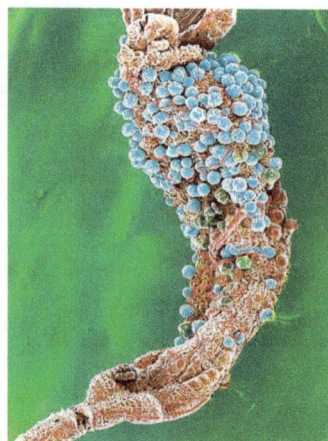

Malaria oocysts on stomach of mosquito.

Coloured TEM of a malarial oocyst in the gut of Anopheles.

When the female *Anopheles* draws a blood meal from an individual infected with malaria, the male and female gametocytes of the parasite find their way into the gut of the mosquito. The molecular and cellular changes in the gametocytes help the parasite to quickly adjust to the insect host from the warm-blooded human host and then to initiate the sporogonic cycle. The male and female gametes fuse in the mosquito gut to form zygotes, which subsequently develop into actively moving ookinetes that burrow into the mosquito midgut wall to develop into oocysts. Growth and division of each oocyst produces thousands of active haploid forms called sporozoites. After the sporogonic phase of 8–15 days, the oocyst bursts and releases sporozoites into the body cavity of the mosquito, from where they travel to and invade the mosquito salivary glands. When the mosquito thus loaded with sporozoites takes another blood meal, the sporozoites get injected from its salivary glands into the human bloodstream, causing malaria infection in the human host. It has been found that the infected mosquito and the parasite mutually benefit each other and thereby promote transmission of the infection. The *Plasmodium*-infected mosquitoes have a better survival and show an increased rate of blood-feeding, particularly from an infected host.

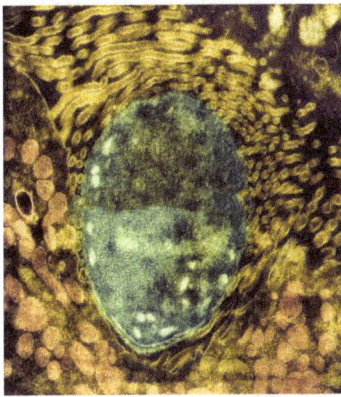

Coloured transmission electron micrograph (TEM) of malaria ookinete in the gut of Anopheles mosquito.

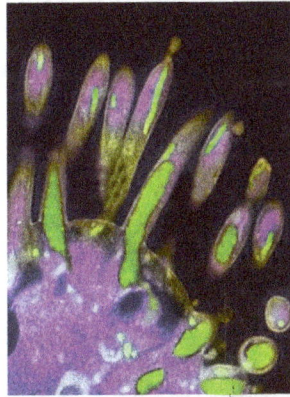

A false-colored electron micrograph showing a malaria sporozoite migrating through the midgut epithelia.

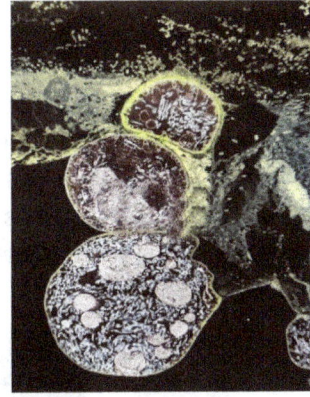

Coloured TEM of malaria oocysts in the gut of Anopheles mosquito.

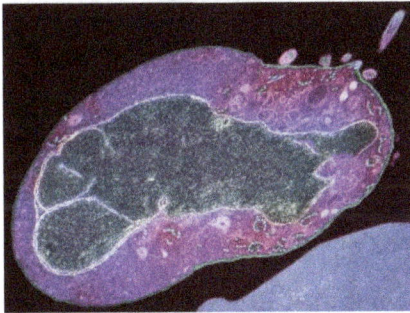

Coloured transmission electron micrograph (TEM) of a male malaria microgametocyte in a Anopheles gut, releasing a male microgamete (purple, at top right).

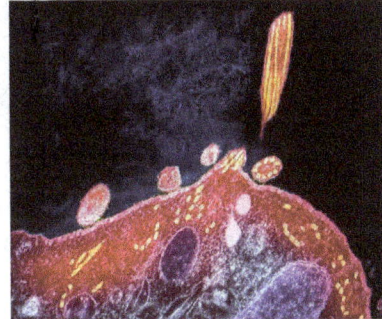

Coloured transmission electron micrograph (TEM) of a male malaria microgametocyte in a Anopheles gut, releasing a male microgamete (at top right).

Schizogony in the Human Host

Man is the intermediate host for malaria, wherein the asexual phase of the life cycle occurs. The sporozoites inoculated by the infested mosquito initiate this phase of the cycle from the liver, and the latter part continues within the red blood cells, which results in the various clinical manifestations of the disease.

Pre-erythrocytic Phase – Schizogony in the Liver

With the mosquito bite, tens to a few hundred invasive sporozoites are introduced into the skin. Following the intradermal deposition, some sporozoites are destroyed by the local macrophages, some enter the lymphatics, and some others find a blood vessel. The sporozoites that enter a lymphatic vessel reach the draining lymph node wherein some of the sporozoites partially develop into exoerythrocytic stages and may also prime the T cells to mount a protective immune response.

The sporozoites that find a blood vessel reach the liver within a few hours. It has recently been shown that the sporozoites travel by a continuous sequence of stick-and-slip motility, using the thrombospondin-related anonymous protein (TRAP) family and an actin–myosin motor. The sporozoites then negotiate through the liver sinusoids, and migrate into a few hepatocytes, and then multiply and grow within parasitophorous vacuoles. Each sporozoite develop into a schizont containing 10,000–30,000 merozoites (or more in case of *P. falciparum*). The growth and development of the parasite in the liver cells is facilitated by a favorable environment created by the circumsporozoite protein of the parasite. The entire pre-eryhrocytic phase lasts about 5–16 days depending on the parasite species: on an average 5-6 days for *P. falciparum*, 8 days for *P. vivax,* 9 days for *P. ovale,* 13 days for *P. malariae* and 8-9 days for *P. knowlesi*. The pre-erythrocytic phase remains a "silent" phase, with little pathology and no symptoms, as only a few hepatocytes are affected. This phase is also a single cycle, unlike the next, erythrocytic stage, which occurs repeatedly.

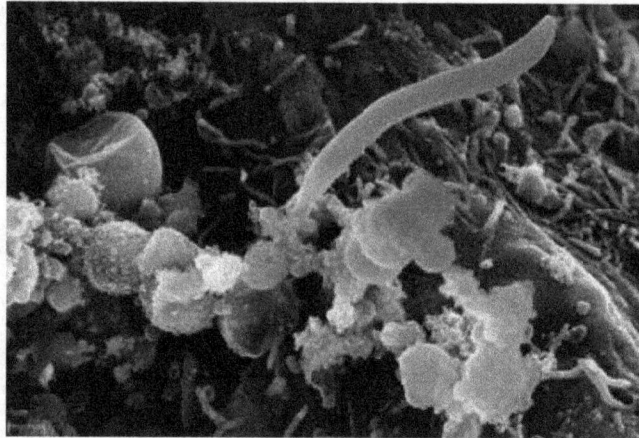

A Plasmodium sporozoite entering a liver cell.

The merozoites that develop within the hepatocyte are contained inside host cell-derived vesicles called merosomes that exit the liver intact, thereby protecting the merozoites from phagocytosis by Kupffer cells. These merozoites are eventually released into the blood stream at the lung capillaries and initiate the blood stage of infection thereon.

In *P. vivax* and *P. ovale* malaria, some of the sporozoites may remain dormant for months within the liver. Termed as hypnozoites, these forms develop into schizonts after some latent period, usually of a few weeks to months. It has been suggested that these late developing hypnozoites are genotypically different from the sporozoites that cause acute infection soon after the inoculation by a mosquito bite, and in many patients cause relapses of the clinical infection after weeks to months.

Erythrocytic Schizogony – Centre Stage in Red Cells

Red blood cells are the 'centre stage' for the asexual development of the malaria parasite. Within the red cells, repeated cycles of parasitic development occur with precise periodicity, and at the end of each cycle, hundreds of fresh daughter parasites are released that invade more number of red cells.

The merozoites released from the liver recognize, attach, and enter the red blood cells (RBCs) by multiple receptor–ligand interactions in as little as 60 seconds. This quick disappearance from the circulation into the red cells minimises the exposure of the antigens on the surface of the parasite, thereby protecting these parasite forms from the host immune response. The invasion of the merozoites into the red cells is facilitated by molecular interactions between distinct ligands on the merozoite and host receptors on the erythrocyte membrane. *P. vivax* invades only Duffy blood group-positive red cells, using the Duffy-binding protein and the reticulocyte homology protein, found mostly on the reticulocytes. The more virulent *P. falciparum* uses several different receptor families and alternate invasion pathways that are highly redundant. Varieties of Duffy binding-like (DBL) homologous proteins and the reticulocyte binding-likehomologous proteins of *P. falciparum* recognize different RBC receptors other than the Duffy blood group or the reticulocyte receptors. Such redundancy is helped by the fact that *P. falciparum* has four Duffy binding-like erythrocyte-binding protein genes, in comparison to only one gene in the DBL-EBP family as in the case of *P. vivax*, allowing *P. falciparum* to invade any red cell.

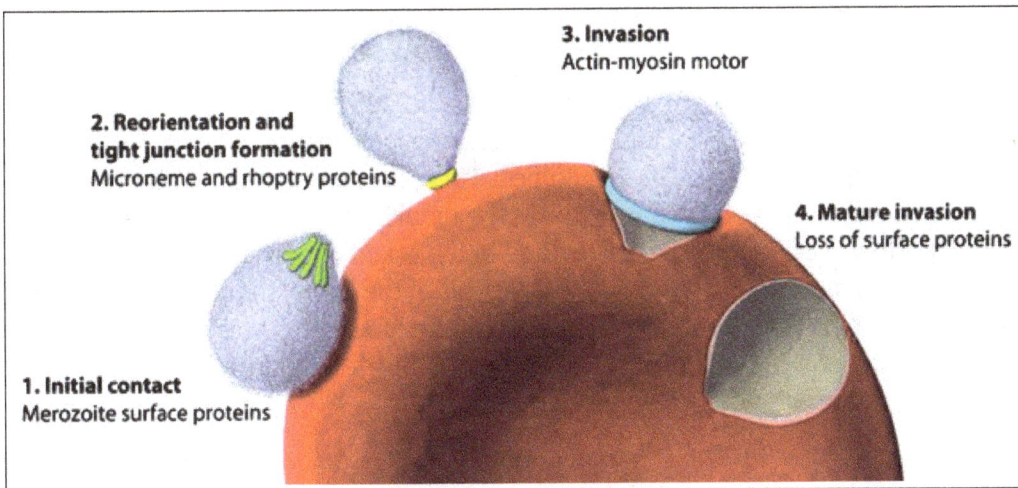

3. Invasion
Actin-myosin motor

2. Reorientation and tight junction formation
Microneme and rhoptry proteins

4. Mature invasion
Loss of surface proteins

1. Initial contact
Merozoite surface proteins

Process of Invasion of Red Cells by Merozoites.

The process of attachment, invasion, and establishment of the merozoite into the red cell is made possible by the specialized apical secretory organelles of the merozoite, called the micronemes, rhoptries, and dense granules. The initial interaction between the parasite and the red cell stimulates a rapid "wave" of deformation across the red cell membrane, leading to the formation of a stable parasite–host cell junction. Following this, the parasite pushes its way through the erythrocyte bilayer with the help of the actin–myosin motor, proteins of the thrombospondin-related anonymous protein family (TRAP) and aldolase, and creates a parasitophorous vacuole to seal itself from the host-cell cytoplasm, thus creating a hospitable environment for its development within the red cell. At this stage, the parasite appears as an intracellular "ring".

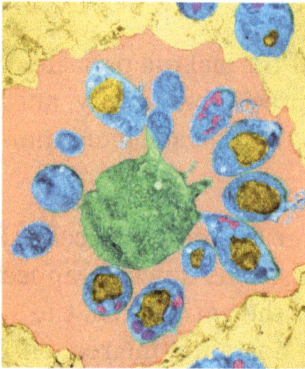

TEM of *P. falciparum* schizont (X2810).

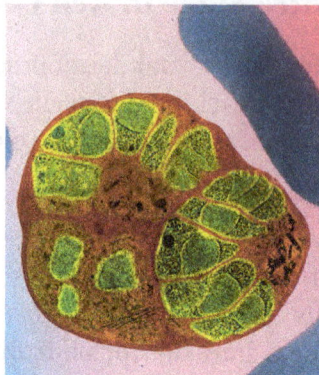

Coloured TEM of a human red blood cell infected with merozoites (green).

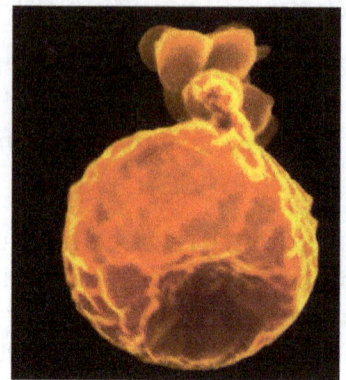

Merozoites of the malaria parasite bursting out of the red blood cell.

Within the red cells, the parasite numbers expand rapidly with a sustained cycling of the parasite population. Even though the red cells provide some immunological advantage to the growing parasite, the lack of standard biosynthetic pathways and intracellular organelles in the red cells tend to create obstacles for the fast-growing intracellular parasites. These impediments are overcome by the growing ring stages by several mechanisms: by restriction of the nutrient to the abundant hemoglobin, by dramatic expansion of the surface area through the formation of a tubovesicular network, and by export of a range of remodeling and virulence factors into the red cell. Hemoglobin from the red cell, the principal nutrient for the growing parasite, is ingested into a food vacuole and degraded. The amino acids thus made available are utilized for protein biosynthesis and the remaining toxic heme is detoxified by heme polymerase and sequestrated as hemozoin (malaria pigment).

The parasite depends on anaerobic glycolysis for energy, utilizing enzymes such as pLDH, plasmodium aldolase etc. As the parasite grows and multiplies within the red cell, the membrane permeability and cytosolic composition of the host cell is modified. These new permeation pathways induced by the parasite in the host cell membrane help not only in the uptake of solutes from the extracellular medium but also in the disposal of metabolic wastes, and in the origin and maintenance of electrochemical ion gradients. At the same time, the premature hemolysis of the highly permeabilized infected red cell is prevented by the excessive ingestion, digestion, and detoxification of the host cell hemoglobin and its discharge out of the infected RBCs through the new permeation pathways, thereby preserving the osmotic stability of the infected red cells.

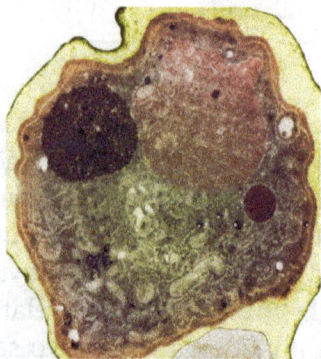

Coloured TEM of a malaria trophozoite in a red blood cell.

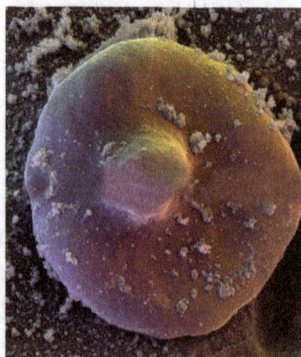

Coloured scanning electron micrograph (SEM) of a malaria infected red blood cell.

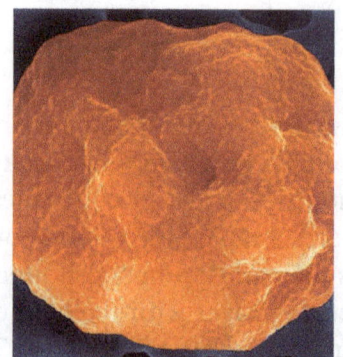

SEM of red cell infected with *P. falciparum* (X4,000).

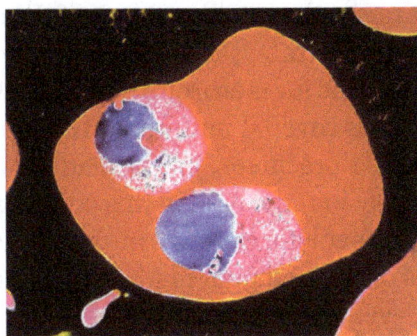

False-colour transmission electron micrograph of two merozoites of *P. falciparum* (blue & pink) parasitising a red blood cell.

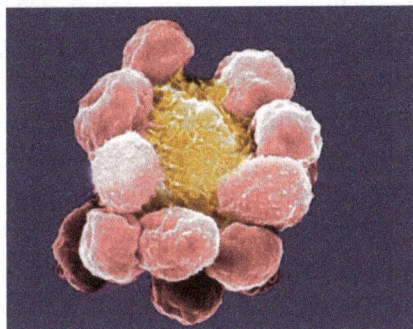

Coloured SEM of a dendritic cell (orange) surrounded by red blood cells infected with *P. falciparum*.

The erythrocytic cycle occurs every 24 hours in case of *P. knowlesi*, 48 h in cases of *P. falciparum*, *P. vivax* and *P. ovale* and 72 h in case of *P. malariae*. During each cycle, each merozoite grows and divides within the vacuole into 8–32 (average 10) fresh merozoites, through the stages of ring, trophozoite, and schizont. At the end of the cycle, the infected red cells rupture, releasing the new merozoites that in turn infect more RBCs. With sunbridled growth, the parasite numbers can rise rapidly to levels as high as 10^{13} per host.

A small proportion of asexual parasites do not undergo schizogony but differentiate into the sexual stage gametocytes. These male or female gametocytes are extracellular and nonpathogenic and help in transmission of the infection to others through the female anopheline mosquitoes, wherein they continue the sexual phase of the parasite's life cycle. Gametocytes of *P. vivax* develop soon after the release of merozoites from the liver, whereas in case of *P. falciparum*, the gametocytes develop much later with peak densities of the sexual stages typically occurring 1 week after peak asexual stage densities.

Host Cell Invasion by Malaria Parasites

The complex life cycle of the malaria parasite includes three specialized invasive stages, distinct both in terms of their cellular architecture and in their choice of target host cell. Despite the dissimilarities between these forms, there are clear parallels in the manner by which they enter their respective host cells.

The development of vaccines that target the invasive stages of the malaria parasite is a major focus of a number of research groups. A novel strategy designed to overcome an immune evasion process mediated by a class of antibodies specific for merozoite surface protein 1 (MSP-1) was described by Tony Holder. The merozoite surface is covered uniformly with a protein complex comprising four polypeptides derived from the MSP-1 precursor, in association with two other proteins encoded by distinct genes. At erythrocyte invasion, the bulk of this complex is released from the merozoite surface as a result of an essential proteolytic cleavage of the single membrane-bound component of the complex. Structural studies of the fragment of MSP-1 (called MSP-1$_{19}$) left on the merozoite surface following this processing event have shown it to be a highly compact molecule consisting of two epidermal growth factor-like modules. MSP-1$_{19}$ can induce an immune response that protects against growth of the blood-stage parasite, and it is the basis of a number of experimental blood-stage malaria vaccines currently under development. Certain antibodies against MSP-1$_{19}$ can prevent proteolytic shedding of the complex, thus preventing erythrocyte invasion, and identifying this as a potentially important mechanism of humoral immunity. However, binding of these antibodies can be blocked by another group of antibodies that recognize adjacent or overlapping epitopes, but which themselves have no effect on MSP-1 processing. Holder and colleagues have proposed that the induction of these 'blocking' antibodies has the effect of reducing the protective efficacy of an anti-MSP-1$_{19}$ antibody response, and is therefore deleterious to the host. Making use of the available structural information, they have set about identifying residues that are crucial to recognition by both processing inhibitory and blocking antibodies. By appropriate modification of these residues, they have engineered mutant forms of recombinant MSP-1$_{19}$ that are no longer recognized by known blocking antibodies, but which retain the structures required for recognition by – and, it is hoped, induction of – processing inhibitory antibodies. These interesting approaches may pave the way to the rational design of MSP-1$_{19}$-based polypeptides that are more effective vaccines than the wild-type molecule.

Despite the immunological evidence of the importance of MSP-1 as a target for protective antibodies, its function on the merozoite surface is unknown. In an effort to begin to address this question by reverse genetic means, Brendan Crabb (University of Melbourne, Australia) discussed the results of a study investigating the requirement for MSP-1 for erythrocytic repli-cation, and the ability of sequence from the MSP-1$_{19}$ domain of the rodent malaria parasite Plasmodium chabaudi to complement the function of the P. falciparum molecule. Transfection and gene-targeting technology was used either to disrupt the P. falciparum gene or to replace precisely the nucleotide sequence encoding all but the first 22 residues of the P. falciparum MSP-1$_{19}$ domain with the corresponding, but highly divergent, P. chabaudi sequence. Disruption of the MSP-1$_{19}$ coding sequence did not result in recovery of viable parasites, suggesting that the intact gene product is essential for efficient erythrocytic growth. Intriguingly, though, analysis of transfected cultures by PCR indicated that MSP-1 'knockouts' might be capable of survival through more than a single cycle of erythrocytic growth. In contrast, parasite clones possessing the chimeric gene could be readily isolated. Surprisingly, although the chimeric gene product was expressed and processed normally, the transgenic parasites exhibited no detectable growth disadvantage in vitro (in cultures containing only human erythrocytes) compared with control counterparts. Growth of the chimeric clone, but not control parasites was significantly inhibited in the presence of antibodies specific for the P. chabaudi MSP-1$_{19}$. These results unambiguously demonstrate that the function of the MSP-1$_{19}$ domain is conserved across species, and raise interesting questions regarding the well-documented sequence conservation of the P. falciparum MSP-1$_{19}$ domain, within which polymorphism has been reproducibly observed at only five of the 96 amino acid residues. If the parasite can accommodate the quite divergent P. chabaudi sequence without apparent ill effect, what are the factors that maintain this remarkably high degree of conservation in the field? One suggestion, put forward by Crabb and colleagues, is that the low immunogenicity of the P. falciparum molecule, in combination with its capacity to induce blocking antibodies, might effectively mean that it is not subject to the immune pressures generally thought to be responsible for selecting for antigenic diversity. They raise the possibility that widespread use of MSP-1$_{19}$ as part of a malaria vaccine in the future might change this situation by providing unprecedented selective pressures on the parasite.

P. knowlesi as a Tractable Experimental Model

Much of our knowledge concerning malaria merozoite ultrastructure and morphological aspects of erythrocyte invasion has derived from use of the simian malaria parasite P. knowlesi, primarily because invasive merozoites of this species can be easily isolated – something that has not been achieved reproducibly with P. falciparum. Alan Thomas and John Barnwell both described some exciting developments in the P. knowlesi model. Stable transformation of P. knowlesi using regulatory DNA sequences from other malaria species to control transgene expression has been previously reported. Thomas detailed two further important steps forward:

(1) The successful adaptation of the species to continuous in vitro culture; and

(2) Targeted integration into the genome of linear, transfected DNA constructs by double crossover homologous recombination.

Double crossover homologous recombination enables non-reversible genome modification by true allelic replacement, an approach that has long been possible with the P. berghei system, but that has yet to be achieved with P. falciparum. Thomas described a number of examples of the use of the technology to disrupt or replace P. knowlesi genes of interest, including homologues of the sporozoite stage-specific thrombospondin-related anonymous protein (TRAP), and apical membrane antigen-1 (AMA-1). The ability to maintain P. knowlesi in vitro – particularly if the parasites can be adapted to growth in human erythrocytes – should, for the first time, allow those researchers who do not have access to rhesus monkey colonies to use this species in studies that require manipulation of invasive merozoites. As a result of these two advances, P. knowlesi may now be considered – like the P. berghei system – a powerful tool for analysis of gene function and expression both in vitro and in vivo, with the additional advantage that it is phylogenetically closely related to P. vivax. Barnwell and colleagues have also used transfection to disrupt two P. knowlesi genes encoding MSP related to MSP-3 of P. falciparum and P. vivax. The absence of an obvious phenotype associated with the knockout parasites was presumably disappointing, but is only one case in a steadily accumulating number of such examples, suggesting that the malaria parasite has evolved substantial redundancy with respect to molecules involved in host cell invasion pathways.

Host-cell-binding Proteins

Barnwell also described the identification of at least two P. falciparum genes encoding merozoite proteins related to the reticulocyte-binding proteins PvRBP-1 and PvRBP-2 of P. vivax[12]. These RBP are expressed in a highly localized manner on the surface of the merozoite apical prominence, and are detectable in culture supernatants in a soluble form that binds to reticulocytes. They are likely to be involved in selective binding to and invasion of reticulocytes. Although neither protein appears to be encoded by a gene family in P. vivax, PvRBP-2 has been previously noted to share some homology with the P. yoelii 235 kDa rhoptry protein (Py235), a red blood cell (RBC)-binding protein encoded by a multigene family of at least 35 members. The two new P. falciparum genes exhibit significant homology with PvRBP-2, and might be expressed in a similar location in the merozoite. In an unscheduled presentation, Peter Preiser (NIMR) talked about recent work on the py235 gene family, which was carried out in collaboration with Laurent Rénia (Institut Co-chin de Génétique Moléculaire, Paris, France), Georges Snounou (Institut Pasteur, Paris, France) and Irène Landau (Muséum National d'Histoire Naturelle, Paris, France). Preiser and colleagues had previously analysed the pattern of transcriptional control of members of this family, and had shown that blood-stage merozoites originating from a single schizont may each express a distinct py235 gene. They now have evidence that py235 mRNA expression continues during the insect stages and pre-erythrocytic stages, but that, rather than multiple transcript types being evident, only a single gene (or subclass of genes) is transcribed. Immediately upon the erythrocytic cycle being re-established by invasion and subsequent replication of released liver-stage merozoites, transcription of multiple genes is again detectable. This, Preiser proposes, suggests that a type of 'reset' mechanism operates as the parasite moves on from the erythrocytic stage of the life cycle into an environment where expression of a protein required to exhibit antigenic variation and/or selection of variant host cell types is not required. The mechanisms regulating this unusual developmental pattern of gene expression will no doubt prove fascinating to uncover.

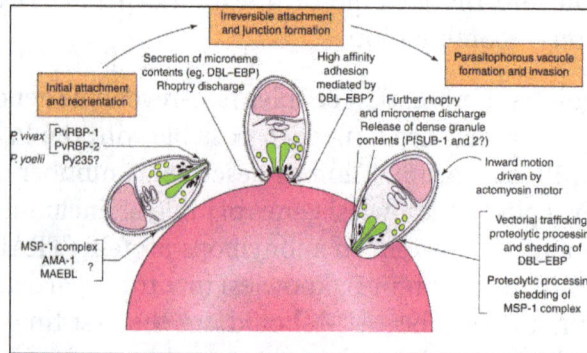

Schematic depicting the early stages of red blood cell (RBC).

In figure, Schematic depicting the early stages of red blood cell (RBC) invasion by the malaria mero-zoite, and the putative roles of the various protein types discussed in the text. The various secretory organelles of the merozoite are shown in dark green (the paired rhoptries), black (micronemes) and pale green (dense granules). The initial, low-affinity interaction with the host cell may involve adhesive ligands stably resident on the merozoite surface such as MSP-1 and other components of the associated protein complex. Alternatively, it may involve AMA-1, which is secreted onto the parasite surface in a truncated form from a primary location in the neck of the rhoptries, or MAEBL. Reorientation of the bound parasite may be favoured by the presence of higher-avidity ligands clustered around the apical prominence, such as PvRBP-1 and -2 in Plasmodium vivax (or functionally equivalent molecules in P. falciparum and P. yoelii). Upon reorientation, tight attach-ment and junction formation may be initiated by exocytosis of micronemal components such as EBA-175 and other DBL–EBPs. The effects of anterior-to posterior trafficking and/or proteolytic shedding of these and other RBC-binding proteins, linked to the action of a sub-pellicular acto-myosin motor, may then aid in driving the parasite into the nascent parasitophorous vacuole. The authors would like to stress that the proposed model is highly speculative. Abbreviations: AMA-1, apical membrane antigen-1; DBL–EBP, Duffy-binding-like erythrocyte-binding protein; EBA-175, 175 kDa erythrocyte-binding antigen; MSP-1, merozoite surface protein-1; PfSUB-1, P. falciparum subtilisin-like protease-1; PvRBP, reticulocyte-binding protein of P. vivax; Py235, P. yoelii 235 kDa rhoptry protein.

The P. vivax and P. knowlesi Duffy-antigen-binding proteins are members of another group of merozoite proteins that function in adhesion to RBC surface receptors. It has become clear that these proteins belong to a large family, referred to as the Duffy-binding-like erythro-cyte-binding protein (DBL–EBP) family, which includes the P. knowlesi β and γ proteins the P. falciparum 175 kDa erythrocyte-binding antigen (EBA-175), and the product of the P. fal-ciparum ebl-1 gene. The extracellular domain of each EBP contains conserved cysteine-rich regions. The N-terminal Cys-rich domains of the EBP – also known as DBL domains – are within region II of the molecule and have been shown to possess receptor-binding activity. Chetan Chitnis described efforts to map regions containing binding residues within DBL do-mains. Chimeric DBL domains containing amino acid stretches from region II of the P. vivax Duffy-binding protein fused to stretches from P. knowlesi β region II were expressed on the surface of mammalian cells and tested for binding to RBCs. The binding residues for the Duffy antigen mapped to the central region between Cys4 and Cys7 of P. vivax region II, and binding residues for sialic acid mapped to a 45 amino acid stretch between Cys4 and Cys5 of P. knowle-si β region II. The adhesive domains of P. vivax Duffy-binding protein and P. falciparum EBA-175, which binds sialic acid residues of glycophorin A, are being developed as vaccines for malaria. However, while P. vivax is completely dependent on the Duffy antigen for RBC inva-sion, P. falciparum is known to invade RBC by multiple pathways. Chitnis described invasion studies with P. falciparum field isolates, which demonstrated that field isolates commonly use multiple RBC invasion pathways that are independent of sialic acid residues of glycophorin A. Such parasites might be expected to be refractory to an anti-EBA-175 immune response. It is interesting, then, that Narum et al. have shown that antibodies specific for region II of EBA-175 have the capacity to interfere with RBC invasion by both sialic-acid-dependent and sialic-acid-independent strains of P. falciparum. The mechanism behind this phenomenon is unclear, but it is proposed to provide continued support for the development of region II as a component of a malaria vaccine. However, the recent demonstration by Reed et al. that targeted disruption of the eba-175 gene results in a stable switch to a sialic-acid-independent invasion phenotype shows clearly that the functional gene product is not required for blood-stage parasite growth in vitro. It is conceivable that immune pressure resulting from use of an EBA-175-based vaccine might select for parasites that do not express EBA-175 and that utilize exclusively alternative invasion pathways.

EBA-175 and the P. knowlesi DBL–EBP appear to be located in merozoite micronemes, but an-other group of type I integral membrane proteins thought to play an adhesive role during RBC invasion are initially sequestered in rhoptries, a morphologically distinguishable set of merozoite secretory organelles. The prototypic member of this family is AMA-1, first identified in P. knowlesi merozoites as the target for an invasion-inhibitory monoclonal antibody. The P. falciparum and P. vivax homologues of AMA-1 are receiving much attention as blood-stage vaccine candidates, and several groups are actively pursuing an understanding of the relationship between the three-di-mensional structure of the molecule, its function, and its capacity to induce protective immune responses. Mick Foley described the use of phage display technology to probe the function of the P. falciparum AMA-1 ectodomain. In an extension of earlier work using phage-derived single-chain variable fragment (scFv) antibodies, Foley has found that panning of random peptide libraries on refolded recombinant AMA-1 can enrich for structures that bind with high affinity to functionally important domains on the molecule, as defined by their ability to interfere with merozoite invasion in vitro, and to compete with the binding of invasion-inhibitory anti-AMA-1 antibodies. In a parallel approach, Foley showed that fine mapping of linear or conformational epitopes recognized by functionally interesting antibodies can be achieved by using the immobilized antibodies either to select mimotopes from the peptide libraries, or to select clones expressing AMA-1 sequences from a library expressing random fragments of the ama-1 gene.

The relatively recently characterized MAEBL family, first identified in the rodent malaria parasites P. yoelii and P. berghei, comprises a set of rhoptry-resident merozoite proteins with chimeric features; the Cys-rich domain lying within the C-terminal part of the MAEBL ectodomain exhibits significant homology to members of the DBL–EBP family, whereas the tandem Cys-rich domains within the N-terminal part of the protein are similar to that of AMA-1. These AMA-1-like Cys-rich domains of MAEBL (called M1 and M2) possess RBC-binding activity, suggesting that they functionally replace the DBL domains within the protein. John Adams presented a description of the P. falciparum single-copy maebl gene and its product. The gene product localizes to the apical complex of merozoites, but it was also detected in a soluble, proteolytically processed form in culture supernatants and was shown to bind to RBC.

Bridging the Gap

Whatever the secretory organelle these host cell-binding proteins are primarily resident in, they all presumably perform their functional role in invasion at the surface of the merozoite. No surface form of any of the DBL–EBP has ever been demonstrated, but relocalization to the merozoite surface has been clearly shown to occur in the case of AMA-1 and MAEBL. Precisely what kind of mechanism could mediate the controlled exocytosis of these integral membrane proteins, and how might they function dynamically during the rapid process of RBC binding, merozoite reorientation and invasion? A recent study might begin to provide some of the answers to this puzzle. The sporozoite protein TRAP is released from micronemes onto the parasite surface, and shed from the sporozoite tail as it glides upon a solid substrate. TRAP is absolutely required for gliding motility, as might be its ookinete paralogue, the circumsporozoite and TRAP-related protein (CTRAP). Kappe et al. used transfection and homologous recombination to introduce a number of modifications into the 3′ region of the P. berghei TRAP gene, encoding the cytoplasmic tail of the protein. Deleting the domain entirely, or substituting certain relatively conserved residues within this domain, did not affect surface expression of TRAP on the transgenic sporozoites, but did have dramatic effects on their gliding phenotype. The absence of the cytoplasmic region abolished motility entirely, whereas appropriate modification of it resulted in sporozoites that did not exhibit the wild-type circular gliding motion but instead moved with a 'pendulum' motion consistent with a defect in release of TRAP at the posterior pole of the sporozoite. The authors suggested that the cytoplasmic domain contains two types of sequences:

(1) Those that act as (or interact with) bridging structures, linking TRAP to a sub-plasmalemmal actomyosin motor; and

(2) Those that control TRAP release following its capping to the posterior pole of the parasite.

The cytoplasmic domain could be functionally complemented by that of MIC2, a homologous protein of the Toxoplasma gondii tachyzoite, previously shown to cap in a myosin-dependent, anterior-to-posterior motion on the parasite surface during invasion. A pattern is beginning to emerge of a class of apicomplexan adherins that – although perhaps sharing little structural similarity between invasive stages – might function in a mechanistically similar manner by mediating binding to host cells via their ectodomains, and inward motion of the parasite during invasion via interaction with a motor system through their cytoplasmic domains.

Proteases and Invasion

The proper functioning of these adhesion proteins might require their appropriate proteolytic shedding as they cap to the posterior of the parasite. MIC2 is subjected to proteolytic processing by at least two parasite proteases as it moves across the surface of the invading T. gondii tachyzoite, and shedding of both the P. knowlesi Duffy-antigen-binding protein17 and the P. yoelii Py235 appears to be the result of proteolytic removal of their transmembrane anchoring domains. Erythrocyte entry by the malaria merozoite is known to require the activity of more than one parasite protease, and these enzymes could present new targets for antimalarial drug development. An improved understanding of the function of these proteases will also shed light on the function of their physiological substrates; in addition to MSP-1 and the host cell-binding proteins discussed above, many other surface and organellar components of the merozoite undergo specific proteolytic processing. Mike Blackman (NIMR) presented recent data on the characterization of two P. falciparum merozoite proteases, both of which localize to organelles within the apical complex. Called P. falciparum subtilisin-like protease-1 and -2 (PfSUB-1 and PfSUB-2), the enzymes belong to the subtilisin-like superfamily of serine proteases, and are subject to a complex post-translational maturation process. PfSUB-1 has been successfully expressed in the baculovirus system in an active form, and its maturation pathway and specificity partially characterized. The role of PfSUB-1 remains unclear, although its preference for cleavage on the C-terminal side of acidic residues means that it is unlikely to be involved in processing and shedding of MSP-1.

Host–Parasite Interactions

Complex Life Cycles within Multiple Hosts

Parasites have evolved elegant strategies to survive and replicate within their hosts. One strategy includes constantly changing their cellular state in order to progress through their life cycle, while simultaneously evading recognition by the host immune system. The vast number of developmental stages, combined with distinct tissue tropisms, increases the complexity of host–parasite interactions. In this topic, we focus on the following species owing to their global impact on human health and influence in the research community:

(1) The Apicomplexans Toxoplasma gondii, which causes toxoplasmosis, Plasmodium spp.,

which cause malaria, and Cryptosporidium spp., which cause the diarrheal disease crypto-sporidosis;

(2) The Kinetoplastids Trypanosoma brucei, which causes African sleeping sickness, Trypano-soma cruzi, which causes Chagas disease, and the Leishmania parasites, which cause both cutaneous and visceral leishmaniasis;

(3) The Diplomonad Giardia lamblia, which causes the intestinal disease giardiasis; and

(4) The Amoebozoa Entamoeba histolytica, which causes amoebic dysentery.

Many of these parasites, such as the Apicomplexans and the Kinetoplastids, are vector-borne, in-tracellular pathogens that complete their life cycle within multiple hosts. One exception is T. bru-cei, which carries out its life cycle extracellularly. Others, such as Cryptosporidium, Entamoeba, and Giardia, can develop into infectious, resistant cysts that survive outside of their hosts and are generally spread via the fecal– oral route.

Because of the important differences in each life-cycle stage, researchers must consider these unique developmental niches as separate systems when studying host–parasite interactions. This is especially important for systems-based analysis, as parasites display periodic stage-dependent gene expression. Accordingly, even slight asynchrony within parasite samples can result in inaccu-rate gene expression measurements, severely limiting statistical power. This is particularly chal-lenging when analyzing clinical samples ex vivo, as parasite populations are rarely homogeneous. Therefore, researchers often utilize specialized techniques in order to synchronize parasites in culture, isolate specific cellular stages from mixed culture, or computationally remove stochastic noise. While this experimental isolation of developmental stages will aid in the understanding of stage-specific host–parasite interactions, it will also be important for future systems-based studies to integrate this knowledge into a multistage model more representative of physiological mixed parasite populations.

Systems Analysis has Advanced our Understanding of Key aspects of Host–Parasite Biology

An enormous amount of information has been produced from the generation of host–parasite systemsbiology datasets within the last decade. The proper integration and interpretation of this 'big data' is critical in order to link experimental findings to useful biological knowledge. Because of the system-wide nature of these datasets, a vast number of important and interesting conclusions can usually be drawn from any given systems-based study, although researchers often choose to pursue only a limited number of noteworthy findings. Interestingly, many sys-tems-based publications have followed up on data that enhance our understanding of specific subfields of host–parasite biology. Although this review will not attempt to encompass all of these findings, we will review some of the leading concepts arising from the analysis of recent genomewide datasets.

Regulation of Parasite Gene Expression

Understanding how parasites regulate gene expression throughout their life cycle within a host is necessary in order to fully appreciate the scope of host–parasite interactions. For example, parasites can actively interfere with host cell translation in order to hijack the cellular resources required for their own gene expression as well as suppress immune responses, as reviewed elsewhere. The system-wide investigation of parasite gene expression is also vital in understanding the coordinated set of events underlying important host– parasite interactions. The upregulation or downregulation of parasite proteins during specific developmental stages is dependent on the host cellular environment and needs to be carefully controlled to ensure parasite survival.

Genome-wide approaches have been particularly important in the elucidation of the regulatory mechanisms governing parasite gene expression in recent years. Although transcriptomics has emerged as a powerful systems-based approach, it must be emphasized that the quantitation of mRNA abundance is often an imperfect indicator of global gene expression. Indeed, for both prokaryotic and eukaryotic organisms, it has been demonstrated that mRNA levels correlate with protein expression for only 50–70% of genes, and for protozoan parasites, this number may be even lower. Systems-based approaches have been especially useful for characterizing the dynamic control of parasite gene expression in recent years. These studies have revealed that precise control of gene expression is essential to drive the dramatic transformation that takes place as parasites cycle through developmental stages, and that the apparent lack of tight transcriptional regulation is remedied by extensive post-transcriptional mechanisms.

In particular, translational delay, a process in which protein expression is actively suspended for expressed mRNA transcripts, is a common strategy employed by many protozoan parasites. Translational delay may be a particularly advantageous strategy for parasites, as they must quickly adapt to new environments and undergo developmental switching in order to survive; storing transcripts necessary for such adaptations allows for rapid changes in gene expression by circumventing the time needed for transcription. Genome-wide next-generation sequencing of both steady-state mRNA as well as polysome-associated transcripts during the asexual erythrocytic stage in P. falciparum revealed widespread translational repression across the genome during different stages of the parasite life cycle. Surprisingly, more than 30% of parasite genes were found to be associated with translational delay. Many of the repressed genes appeared to be regulated by cell cycle stage and they clustered into discrete biological processes. For example, many genes associated with early-stage processes, such as nutrient acquisition and erythrocyte remodeling, were transcribed during the trophozoite or schizont stages, and were only actively translated immediately following merozoite invasion. Another genome-wide ribosomal profiling study of P. falciparum blood stages provides additional support for a model whereby transcription of important merozoite genes occurs during the previous stage and is translationally upregulated during invasion. Translational delay has also been demonstrated during the sexual gametocyte stage by temporary storage of specific transcripts in P-bodies. Unlike the majority of other eukaryotic organisms, Trypanosomes transcribe almost all of their genes as large polycistronic clusters, and thus lack transcriptional control for most genes. Despite the absence of regulation at the level of transcription, transcriptomic surveys have revealed extensive variation in mRNA abundance across developmental stages, suggesting widespread posttranscriptional control. Furthermore, the comparison of proteomic expression using SILAC and MS to transcriptomic datasets suggests that like Plasmodium, mRNA abundance does not predict protein expression for at least 30% of the T. brucei genome. The integration of data surveying global protein expression, polysome-associated transcript abundance, and total mRNA during these stages revealed extensive translational repression during the time when T. brucei prepares for transmission.

Parasite Utilization of Host Resources

While eukaryotic pathogens are often able to synthesize a number of nutrients required for growth de novo, it is often more advantageous to conserve the energy required for biosynthesis and to instead hijack hostderived resources. This is especially true for the acquisition of host lipids, as protozoan parasites must quickly assemble a large amount of new membrane during replication within host cells. In Apicomplexan parasites such as P. falciparum and T. gondii, fatty acids are taken up from the host and converted into triacylglycerides, where they are then stored in lipid bodies. Recently, a system-wide survey of the Plasmodium lipidome during liver-stage infection revealed a significant enrichment in fatty acids important for membrane biogenesis, including phosphatidylcholine. Upon further investigation, it was found that the parasite actively acquires host-derived phosphatidylcholine and that this process is essential for parasite survival within hepatocytes. It has also been shown that Leishmania parasites while unable to synthesize sphingomyelin themselves, are able to hydrolyze host sphingomyelin in order to produce essential metabolites. A comparative genomics study identified a parasite enzyme, LaISCL, which is responsible for the degradation of host-derived sphingomyelin, and showed that this process is necessary for the proliferation of L. major parasites within their mammalian hosts. More recently, the same group showed that this enzyme is also responsible for sphingomyelin turnover in Leishmania amazonensis, although in this species, the role of sphingomyelin degradation in promoting virulence is quite different.

Many protozoan parasites live an intracellular auxotrophic lifestyle, actively acquiring metabolites from their nutrient-rich host in order to survive. For instance, blood-stage Plasmodium parasites have lost the ability to biosynthesize purine rings or amino acids, and therefore scavenge host nucleotides to synthesize DNA and catabolize host hemoglobin to generate amino acids. Recent system-wide metabolomic studies have been instrumental in profiling the complex exchange of nutrients between parasites and their hosts. A comprehensive MS-based approach revealed significant modulation of host metabolites during blood-stage Plasmodium development. The authors found that host arginine depletion was particularly extensive, suggesting that this may contribute to human malarial hypoargininemia and progression to cerebral malaria. Another Apicomplexan parasite, T. gondii, relies on host nutrients, such as carbon, in order to proliferate within host cell vacuoles. In a combined metabolomic and stable isotope labeling approach, a recent study mapped the carbon metabolism pathway for T. gondii tachyzoites. This systemsbased analysis revealed that active catabolism of host glucose and glutamine through an oxidative TCA cycle is essential for parasite replication. Through these and similar systems-biology-based surveys, it is becoming clear that protozoan parasites have evolved complex strategies to both usurp and exploit host resources.

Host Immune Response to Parasitic Infection

In order to fully appreciate the complexity of host– parasite interactions, the host immune response must be considered. It is well established that while most protozoan infections are self-limiting in immunocompetent hosts, however, immunocompromised individuals can develop severe and often life-threatening disease, suggesting that an effective immune response is essential for regulating parasitic disease. Many omic-based strategies have contributed to our current knowledge of how the innate and adaptive immune systems resist parasitic infection, and in many cases, exacerbate disease. In particular, recent transcriptomic analyses of host–parasite systems have implicated the host innate Toll-like receptor (TLR) and interferon (IFN)-mediated proinflammatory pathways in the regulation of disease progression. Microarray analysis of malaria patient samples demonstrated an upregulation of TLR signaling genes that had sites for IFNinducible transcription factors. Upon subsequent analysis of Plasmodium-infected rodents, it was revealed that TLR9 and MyD88 are critical to initiate the cytokine responses leading to acute malaria in vivo. Another transcriptomic analysis of patient responses further confirmed the enhancement of IFN-stimulated genes (ISGs) upon infection with malaria parasites, and interestingly, the same study determined that TLR9-independent sensing of AT-rich Plasmodium DNA induces type I IFNs. In a dual RNASeq approach, a recent report mapped host and pathogen transcriptomes during acute and chronic infection with T. gondii. Analysis of the differentially expressed transcripts revealed that many of the acute infectionspecific genes included ISGs such as guanylate-binding proteins. Chronic infection-specific transcripts were shown to comprise a unique set of immune genes, including those important for antigen recognition and presentation. Thus, these systems-level analyses indicate that innate sensing of protozoan pathogens is important for the induction of proinflammatory responses aimed at controlling infection.

Parasitic disease is an evolutionary arms race; as our immune systems attempt to fight off infection, pathogens quickly respond by adapting to and subverting these attacks, often through elegant biological maneuvers. Multiple -omic-based surveys have contributed to our knowledge of how protozoan parasites actively manipulate the host immune response in order to avoid detection. Over a decade of systems-biology research has shown that T. gondii downregulates the innate immune response by multiple mechanisms. This includes preventing host nuclear translocation of proinflammatory transcription factors such as nuclear factor kappa β (NF-κβ) and signal transducer and activator of transcription 1 (STAT1α), as well as upregulating anti-inflammatory pathways such as those involving the suppressor of cytokine signaling (SOCS) proteins.

A notable systems-based study utilized transcriptomics and pathway analysis to show that Toxoplasma actively regulates host immune responses, and through forward genetics, discovered a parasite rhoptry kinase, ROP16, that is secreted into the host cytoplasm to interfere with STAT signaling. Additionally, Plasmodium parasites also secrete virulence factors that specifically block host innate immune signaling. During liver-stage development, Plasmodium circumsporozoite protein (CSP) is exported and localized to the host cell nucleus where it interferes with the nuclear translocation of NF-κβ, and microarray analysis confirms that at least 40 NF-κβ-responsive genes are downregulated with CSP expression. Likewise, in the blood stages of the parasite, a highthroughput protein interaction screen found that Plasmodium merozoite surface protein 1 (MSP1) specifically binds to the human proinflammatory cytokine S100P, and that this interaction blocked activation of the host NF-κβ-mediated innate immune response. Through these and other genome-wide investigations, it is clear that while the host innate immune system is essential in controlling parasitic infection, parasites have evolved complex strategies to effectively dampen these responses.

Parasite Damage and Host Responses

The life cycle of *Plasmodium* involves an alternation between mosquito and vertebrate hosts. A mosquito acquires the parasite when it feeds upon the blood of an infected animal. *Plasmodium* undergoes sexual reproduction in the gut of the mosquito; the offspring then migrate to the insect's salivary glands, and are transferred into a vertebrate host when the mosquito feeds. In vertebrates, the parasites undergo successive rounds of asexual multiplication, first inside the host's liver cells and later in red blood cells.

In the mosquito, *Plasmodium* is fairly benign, but in its vertebrate hosts it causes severe illness, characterised by periodic bouts of intense fever. Malaria caused by *P. falciparum*, one of the four *Plasmodium* species that infect humans, can produce life-threatening complications such as cerebral malaria (in which the parasite affects the brain), renal failure and pulmonary oedema. Malaria is also associated with conditions such as anaemia and hypoglycaemia, which can have a severe impact upon health and survival, particularly among children and pregnant women.

The symptoms of malaria are the result of complex interactions between *Plasmodium* and the immune system of its host. Although the biology of the parasite is now well understood, the exact mechanisms by which it causes disease are the subject of extensive controversy and ongoing research. The main processes that are believed to be involved are outlined in this essay, and illustrated in figure.

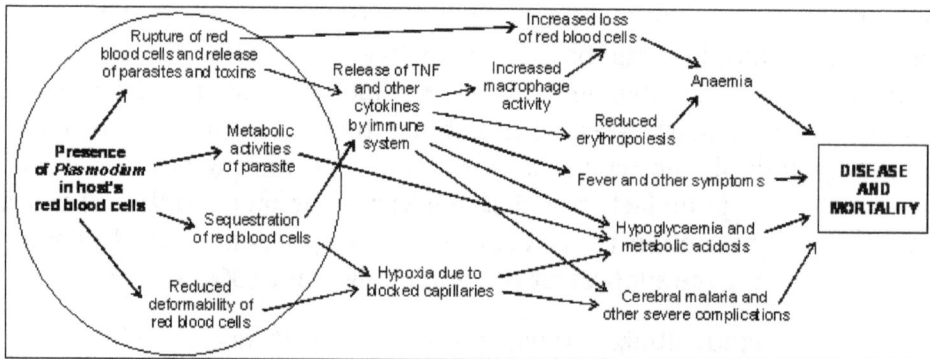

Possible mechanisms by which *Plasmodium* causes disease.

The Immune Response

The periodic bouts of fever that occur in malaria coincide with the synchronised rupture of *Plasmodium*-infected red blood cells. This causes the release of parasites *en masse* into the blood stream, along with pigments and toxins that have accumulated inside the red blood cells as a result of the parasites' metabolic activities. The presence of large quantities of parasite material in the blood triggers a dramatic immune response, mediated by the secretion of cytokine molecules by the cells of the immune system. Some cytokines—such as 'tumour necrosis factor' (TNF), interferon-gamma, interleukin-12 and interleukin-18—enhance the immune response, stimulating macrophages and other immune cells to destroy parasites by phagocytosis and by the production of toxins. Other cytokines—including interleukin-4, interleukin-10 and TGF-beta—help to regulate the immune response by dampening these effects.

Although the immune response stimulated by cytokines undoubtedly plays an important role in suppressing and killing malaria parasites within the body, excessive production of cytokines can have pathological consequences. The high levels of TNF and other 'inflammatory' cytokines produced in response to *Plasmodium* are responsible for the intense fever associated with malaria. *Plasmodium* triggers the same immune pathways as a variety of other parasites, which may explain why the symptoms of malaria closely resemble those of many other diseases. It is not known whether malarial fever is an adaptive response by the host that helps to kill the parasites, or a pathological reaction caused by over-stimulation of the immune system. Overproduction of inflammatory cytokines may also be responsible for the life-threatening complications such as cerebral malaria that occur in a small proportion of malaria patients.

Sequestration

Red blood cells infected with *Plasmodium falciparum* display protein-rich 'knobs' on their outer surfaces, which cause the cells to adhere to one another and to capillary walls. This adhesion allows parasite-infected cells to remain 'sequestered' in particular organs rather than circulating freely in the bloodstream, helping the parasite to evade the host's immune system. Since almost all of the human deaths attributed to malaria are caused by *P. falciparum*—the only human-infecting *Plasmodium* species with the ability to sequester—it is widely believed that that sequestration plays a key role in cerebral malaria and other fatal complications of the disease. However, there are conflicting theories about how this occurs.

The traditional explanation for cerebral malaria is that sequestration, perhaps combined with the reduction in the deformability of red blood cells that occurs when the cells are infected with *Plasmodium*, leads to the blockage of capillaries in the brain, depriving the tissue of oxygen. However, measurements made using Near Infrared Spectroscopy and Doppler sonography show that levels of blood flow in the brains of cerebral malaria patients are not abnormally low, and individuals who recover from cerebral malaria do not generally exhibit the permanent brain damage that is typically associated with acute oxygen deprivation. It has therefore been suggested that sequestered cells infected with *P. falciparum* harm the brain by causing an excessive immune reaction there, rather than by physically blocking capillaries. Nitric oxide (NO), a substance that is manufactured by macrophages to kill parasites but is also toxic to host cells at high concentrations, has been implicated in this damaging immune reaction. It has been reported that a toxin produced by *P. falciparum* can induce the NO-synthesising enzyme iNOS in human endothelial cells, and iNOS has been found in samples of brain tissue taken during autopsies of cerebral malaria victims.

In pregnant women, *P. falciparum* frequently sequesters in the placenta, where rich capillary beds and weakened immune responses create a hospitable environment for the parasite. Placental malaria can have harmful consequences for the foetus, disrupting its supply of oxygen and nutrients, and increasing the risk of premature delivery. These problems cause babies born to malaria-infected mothers to have an unusually high probability of low birth weight, which in turn is associated with higher levels of infant mortality.

Anaemia

Plasmodium, like many parasites, can damage its host by causing anaemia—a reduction in the ability of the blood to transport oxygen, which leads to lethargy and (in very extreme cases) can be fatal. The decrease in red blood cell concentration that is responsible for malarial anaemia occurs both through an increase in the rate at which red blood cells are destroyed and a decrease in the rate at which new ones are produced. *Plasmodium* not only causes the rupture of parasitized red blood cells, but stimulates the activity of macrophages in the spleen, which then destroy both parasitized and unparasitized red blood cells. (During malaria infection, unparasitized red blood cells may be targeted because they have abnormally rigid membranes, or because malarial antigens present in the bloodstream bind to their surfaces.) TNF-alpha and other cytokines produced during malaria depress the rate of erythropoiesis (the manufacture of new red blood cells), further contributing to anaemia. The health risks due to malaria-related anaemia are particularly severe in pregnant women, and there is some evidence that maternal anaemia is associated with anaemia in the foetus, which increases the risk of infant mortality.

Physiological Changes

Parasites can produce physiological abnormalities in their host, which may have harmful consequences throughout the body, not just in the tissues or organs in which the parasite is present. Malaria patients commonly exhibit hypoglycaemia—a particular problem in pregnant women, in whom hypoglycaemia may be a cause of low foetal birth weight—and metabolic acidosis, which is the cause of a significant number of malaria-related fatalities among young children. Possible explanations for these physiological changes include anaerobic consumption of glucose by the parasites and hypoxia due to the blockage of capillaries by parasite-infected red blood cells. There is also strong evidence that the cytokines produced during malaria can cause hypoglycaemia and acidosis by inducing changes in the body's carbohydrate metabolism. Hypoglycaemia and acidosis can be induced in the absence of malaria by injecting TNF into animals, and in human malaria patients TNF levels correlate with hypoglycaemia.

Factors affecting the Severity of Disease

A given variety of parasite will not affect all hosts in an identical way. The likelihood of parasite infection being established and the severity of the damage caused depend upon many factors, included the host's genotype, age, nutritional status and immunological history.

It has been known for a long time that individuals with certain genotypes, such as those carrying the famous 'sickle-cell' haemoglobin allele, are protected against infection by *Plasmodium*. More recently, researchers have also discovered genetic variations among humans that affect the severity of the damage caused when *Plasmodium* infection does become established. For example, a particular allele of the TNF-alpha gene-promoter region was found to be associated with a high probability of developing fatal cerebral malaria in patients infected with *P. falciparum*.

The outcome of an infection also depends upon the genotype of the parasite. An obvious (albeit somewhat unnatural) illustration of this is provided by the existence of localised varieties of *P. falciparum* that show an inherited resistance to certain anti-malarial drugs.

People living in areas where malaria is highly endemic, who are repeatedly infected with *Plasmodium*, eventually attain a 'semi-immune' state in which an uneasy balance is reached between the parasite and the host's immune system. Such people continue to harbour significant levels of *Plasmodium* in their blood, but do not show obvious symptoms of disease. Children do not normally acquire this semi-immunity until around the age of five, which partly explains why young children account for the vast majority of deaths that occur due to malaria. Factors that weaken the immune system—such as pregnancy, micronutrient malnutrition, and HIV infection—have been associated with an increased incidence of *Plasmodium* infection, and with a resurgence of symptoms in those chronically infected with the parasite.

Damage Control in Host–Pathogen Interactions

When facing a problem, we would ideally like to be able to eliminate its source. If that is not feasible, we focus our efforts on damage control, by attempting to minimize the consequences of the problem at hand. Likewise, host defense from infections can employ two distinct strategies of protection: one aims to reduce or eliminate the invading pathogen, whereas the other reduces the damage to the host inflicted by a given pathogen burden. The two strategies are referred to as resistance and tolerance, respectively. Resistance mechanisms are generally mediated by the immune system. Mechanisms of tolerance are far less well understood, particularly in animal hosts. The report by Seixas et al. in this issue of PNAS describes a mechanism of host tolerance used during infection with Plasmodium parasites. This study provides one of the first insights into molecular mechanisms of infection tolerance in mammals and is likely to be applicable, at least in principle, to a broad range of infectious diseases.

Resistance and tolerance have long been recognized as distinct host defense strategies used by plants to deal with their pests. The distinction between the two strategies is fundamentally important, because they rely on discrete molecular mechanisms and entail different consequences for the evolutionary dynamics of host–pathogen interactions. Surprisingly, the concepts of resistance and tolerance are seldom applied to studies of animal immunity, where research has traditionally focused almost exclusively on mechanisms of resistance. Consequently, very little is known about the underlying mechanisms of tolerance and their utility for host protection. Tolerance mechanisms do exist in animals, however, as documented by the few studies reported to date: Råberg et al. described a variation in tolerance to Plasmodium infection in several inbred mouse strains, which was probably the first clear description of tolerance in animals, and Schneider and colleagues in a series of elegant studies dissected the contributions of resistance and tolerance to Drosophila defense from infections. Together, these studies have indicated the existence of powerful but poorly understood defense mechanisms that contribute to host survival from infections. Clearly, this line of research will have to be continued and significantly expanded to reveal what is likely to be a plethora of mechanisms that contribute to host tolerance to infections.

First, the symptoms of infectious diseases can be either due to direct damage to the host inflicted by the pathogen, or due to immunopathology—the collateral damage to the host tissues caused by the immune response. Accordingly, tolerance mechanisms may be broadly divided into mechanisms that protect the host tissues from direct pathogen-induced damage, for example, caused by microbial toxins, and the mechanisms that limit immunopathology. The latter in turn can be subdivided into immunoregulatory mechanisms, which control the intensity and duration of the immune and inflammatory responses, and the mechanisms that render tissues more resistant to inflammatory damage. Second, the relative contribution of the two types of mechanisms most likely depends on the replication, transmission, and host adaptation strategies of microorganisms. An extreme example illustrating this point is the nature of host–commensal interactions. Here, the host tolerates microbial colonization of enormous proportions without any adverse effects (in fact, with a number of essential benefits). In large part, this is due to potent immunoregulatory mechanisms that maintain a host–commensal homeostasis at the sites of colonization, such as the colon. If the normal immunoregulatory state is disrupted, however, the ensuing problems are due almost entirely to the inappropriate immune and inflammatory responses triggered by the commensal microbes.

Most pathogenic infections will elicit both types of tissue damage, and therefore the host may commonly employ both types of tolerance mechanisms. The case in point is infection with Plasmodium, the causative agent of malaria. This protozoan parasite can cause direct damage to the host through hemolysis and anemia, and indirect damage attributable to immunopathology. The latter is responsible for some forms of severe malaria, including cerebral malaria, which is caused by the disruption of the blood–brain barrier followed by the often lethal inflammatory response in the brain. During the blood stage of infection, Plasmodium invades red blood cells (RBCs), where it degrades hemoglobin, resulting in the release of free heme. Because free heme is toxic to the host and to the parasite, both try to convert it to more innocuous chemical forms. Plasmodium converts heme into a polymer called hemozoin, which is nontoxic to the parasite. This conversion can be blocked by chloroquine, which presumably explains its therapeutic properties in malaria patients and the popularity of gin and tonic among European settlers in the parts of the world where malaria is common. Circulating heme has pro-inflammatory properties: it induces reactive oxygen species (ROS) and activates Toll-like receptor 4 on macrophages. Heme detoxification occurs through the induction of heme oxygenase-1 (HO-1), which converts it into biliverdin, carbon monoxide (CO), and iron. Biliverdin is further metabolized into bilirubin, which is excreted in bile. Importantly, HO-1 has previously been found to play an essential role in protection from the experimental cerebral malaria by preventing heme-induced destabilization of the blood–brain barrier.

In the new study, Seixas et al. demonstrate that the heme detoxification activity of HO-1 is a critical component of host tolerance to Plasmodium infection-associated liver failure. Infection with *Plasmodium chabaudi chabaudi* (Pcc) resulted in hepatic failure in susceptible mouse strains due to extensive hepatocyte apoptosis caused by tumor necrosis factor (TNF) produced during the infection. Expression of HO-1 determined susceptibility to this form of severe malaria, as demonstrated by the analyses of HO-1-deficient mice. Free heme was found to sensitize hepatocytes for TNF-induced apoptosis through a mechanism involving ROS generation. Degradation of heme by HO-1 was required to prevent Pcc induced hepatic failure and mortality. Importantly, this protective activity of HO-1 was found to have no effect on parasite burden, thus clearly demonstrating its key contribution to host tolerance without affecting host resistance to Plasmodium infection. Remarkably, Seixas et al. found this tolerance mechanism to be extremely potent, as it could afford a complete protection from Pcc infection-associated mortality. Furthermore, elimination of heme-induced ROS by administration of a pharmacological antioxidant, *N*-acetylcysteine (NAC), had a potent therapeutic effect, even when NAC was administered 4 days after infection. Importantly, although administration of NAC protected the otherwise susceptible mice from Pcc-induced mortality, it had no effect on parasite burden. This finding thus not only offers an attractive therapeutic strategy but also provides an interesting example of a pharmacological manipulation of host tolerance. A more familiar example of the same strategy is the use of cyclooxygenase inhibitors (such as aspirin or ibuprofen) during influenza virus infections. The fever-lowering effect of these drugs enhances host tolerance and presumably has no direct effect on the virus. It could be expected that blocking different aspects of the inflammatory response is likely to have a positive effect on host tolerance in a variety of infectious diseases, especially when boosting the antimicrobial activity is not an option, or when the presence of the pathogen is easily tolerated in the absence of immunopathology.

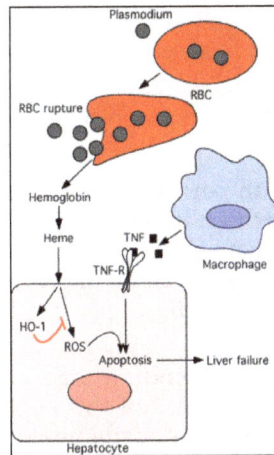

Role of HO-1 in cytoprotection during Plasmodium infection.

In figure above, Plasmodium infection results in lysis of red blood cells (RBC) followed by release of hemoglobin and free heme. Free heme triggers production of reactive oxygen species (ROS). ROS sensitizes hepatocytes to undergo apoptosis in response to tumor necrosis factor (TNF) produced during infection. Heme oxygenase-1 (HO-1) is induced in hepatocytes and degrades free heme, thus protecting hepatocytes from apoptosis. HO-1 thus plays a critical role in tissue protection during Plasmodium infection.

The study by Seixas et al. has identified heme-induced ROS as a major component of host susceptibility to *Pcc* infection-associated liver damage. Excessive ROS levels are likely to play an important role in determining host tolerance in a variety of infectious diseases, particularly when an excessive inflammatory response is a limiting factor. Although immunoregulatory mechanisms that limit the excessive immune and inflammatory responses have been studied extensively, the mechanisms that render tissues resistant to immune and inflammatory damage are poorly understood. The importance of the study by Seixas et al. is that it describes just such a mechanism, which operates in the context of infection with a deadly parasite.

References

- SixEssentialAspects: cbu.edu, Retrieved 5 February, 2019

- Effects-of-parasitic-life-on-parasites-and-host, protozoa, invertebrate-zoology: biologydiscussion.com, Retrieved 13 April, 2019

- Life-cycle: malariasite.com, Retrieved 9 January, 2019

- Malaria: andrewgray.com, Retrieved 2 March, 2019

Permissions

All chapters in this book are published with permission under the Creative Commons Attribution Share Alike License or equivalent. Every chapter published in this book has been scrutinized by our experts. Their significance has been extensively debated. The topics covered herein carry significant information for a comprehensive understanding. They may even be implemented as practical applications or may be referred to as a beginning point for further studies.

We would like to thank the editorial team for lending their expertise to make the book truly unique. They have played a crucial role in the development of this book. Without their invaluable contributions this book wouldn't have been possible. They have made vital efforts to compile up to date information on the varied aspects of this subject to make this book a valuable addition to the collection of many professionals and students.

This book was conceptualized with the vision of imparting up-to-date and integrated information in this field. To ensure the same, a matchless editorial board was set up. Every individual on the board went through rigorous rounds of assessment to prove their worth. After which they invested a large part of their time researching and compiling the most relevant data for our readers.

The editorial board has been involved in producing this book since its inception. They have spent rigorous hours researching and exploring the diverse topics which have resulted in the successful publishing of this book. They have passed on their knowledge of decades through this book. To expedite this challenging task, the publisher supported the team at every step. A small team of assistant editors was also appointed to further simplify the editing procedure and attain best results for the readers.

Apart from the editorial board, the designing team has also invested a significant amount of their time in understanding the subject and creating the most relevant covers. They scrutinized every image to scout for the most suitable representation of the subject and create an appropriate cover for the book.

The publishing team has been an ardent support to the editorial, designing and production team. Their endless efforts to recruit the best for this project, has resulted in the accomplishment of this book. They are a veteran in the field of academics and their pool of knowledge is as vast as their experience in printing. Their expertise and guidance has proved useful at every step. Their uncompromising quality standards have made this book an exceptional effort. Their encouragement from time to time has been an inspiration for everyone.

The publisher and the editorial board hope that this book will prove to be a valuable piece of knowledge for students, practitioners and scholars across the globe.

Index

www.ingramcontent.com/pod-product-compliance
Lightning Source LLC
Chambersburg PA
CBHW082016190326
41458CB00010B/3208

* 9 7 8 1 6 4 7 4 0 0 1 3 2 *